Lecture Notes in Mathematics

Edited by A. Dold and B. Eckmann

Series: Department of Mathematics, University of Maryland, College Park
Adviser: L. Greenberg

478

T0186099

Glenn Schober

Univalent Functions - Selected Topics

Springer-Verlag
Berlin · Heidelberg · New York 1975

Author
Prof. Glenn Schober
Department of Mathematics
Indiana University
Swain Hall East
Bloomington, Indiana 47401
U.S.A.

Library of Congress Cataloging in Publication Data

Schober, Glenn, 1938-
Univalent functions--selected topics.

(Lecture notes in mathematics ; 478)
Bibliography: p.
Includes index.
1. Univalent functions. I. Title. II. Series.
QA3.L28 vol. 478 [QA331] 510'.8s [515'.253]
ISBN 978-3-540-07391-8 75-23099

AMS Subject Classifications (1970): 30A32, 30A36, 30A38, 30A40, 30A60

ISBN 978-3-540-07391-8 Springer-Verlag Berlin · Heidelberg · New York
ISBN 978-0-387-07391-0 Springer-Verlag New York · Heidelberg · Berlin

Offsetdruck: Julius Beltz, Hemsbach/Bergstr.

PREFACE

These notes are from lectures given by the author in 1973-74 at the University of Maryland, during its special year in complex analysis. They are an attempt to bring together some basic ideas, some new results, and some old results from a new point of view, in the theory of univalent functions.

There are really two points of view that are used and intertwined in these notes. The first is to utilize a linear space framework to study sets of univalent functions as they are situated in a space of analytic functions. For example, in Chapter 7 we are interested in compactness of families of univalent functions that lie in the intersection of two hyperplanes, and in Chapter 8 we are interested in their geometry in the sense of convexity theory.

In the same spirit, we consider in Chapter 2 many of the special families of univalent functions and determine the extreme points of their closed convex hulls. This point of view seems to simplify and unify the study of their properties, for example, in solving linear extremal problems. In keeping with this point of view we give in Chapter 1 a derivation of the Herglotz representation based on Choquet's theorem. In this case the route is less elementary, but it serves to establish our point of view.

The second point of view is to study extremal problems using variational considerations. In the absence of a structural formula for a class of functions, variational methods are a very powerful tool. In Appendix C we include the boundary variation from the fundamental work of M. Schiffer, and in Chapters 10 and 11 we apply it to solve some accessible problems and to give geometric properties of solutions to others.

Variational considerations can also be used to study quasiconformal mappings. Our treatment of quasiconformal mappings is not thorough. However, in Chapter 13 we give a very general variational procedure for families of quasiconformal mappings. Some applications are given in Chapter 14. At present, this appears to be a rapidly developing area.

A number of other topics that seem to fit in are included, e.g., the affirmative solution of the Pólya-Schoenberg conjecture, representation of continuous linear functionals, Faber polynomials, properties of quasiconformal mappings, and quasiconformal extensions of univalent functions.

One final comment about the structure of these notes: There are two kinds of problems in the text, those called exercises and those called problems. There is a distinction; namely, the author knows how to solve only the exercises. This comment was delayed until now in the hope that the reader might overlook it, and might go on to solve some of the problems.

Finally, the author wishes to acknowledge the work of Mr. Tom Whitehurst, who proofread these pages, and of Miss Julie Palmer and Mrs. Karen Barker, who typed them.

Bloomington, Indiana — Glenn Schober

December 1974

CONTENTS

CHAPTER 1. Functions with positive real part

Let D be a domain in the complex plane $\mathbb{C}$. Denote by $H(D)$ the set of analytic functions on D. Endowed with the topology of uniform convergence on compact sets, $H(D)$ is a <u>linear topological space</u>. Let $\{K_n\}$ be an exhaustion of D by compact sets, i.e., $K_n \subset K_{n+1}$ and $\bigcup_{n=1}^{\infty} K_n = D$, and for $g \in H(D)$ let $B_n(g) = \{f \in H(D) : |f-g| < 1/n \text{ on } K_n\}$. Then $\{B_n(g) : g \in H(D),\ n = 1,2,\ldots\}$ forms a base for the topology on $H(D)$. Since each $B_n(g)$ is a convex set, $H(D)$ is a <u>locally convex</u> space. The topology is <u>metrizable</u> with metric

$$d(f,g) = \sum_{n=1}^{\infty} 2^{-n} \sup_{K_n} |f-g|/(1+|f-g|) \ .$$

Denote by $H_u(D)$ the set of univalent functions in $H(D)$.

We shall study subsets of $H(D)$ from the point of view of convexity theory. The relevant facts are contained in Appendix A. For the special case of the open unit disk

$$U = \{z : |z| < 1\}$$

we define

$$P = \{f \in H(U) : \operatorname{Re} f > 0 \text{ and } f(0) = 1\} \ .$$

Observe that P is a convex set.

The rest of this chapter is devoted to obtaining the correspondence between P and the set of Borel measures on the unit circle ∂U, that is known as the Herglotz representation. Our method will be to determine the extreme points of P following a recent article of F. Holland [H8] and to deduce the correspondence from Choquet's theorem (Appendix A). This method is neither constructive nor as elementary as customary developments. However, we present it in order to establish a point of view.

THEOREM 1.1 (subordination principle). Suppose $f \in H(U)$, $F \in H_u(U)$, $f(0) = F(0)$, and $f(U) \subset F(U)$. (Write $f \prec F$, f is subordinate to F.) Then

(a) $f = F \circ \omega$ where $\omega \in H(U)$ and $|\omega(z)| \le |z|$,

(b) $|f'(0)| \le |F'(0)|$ with equality iff $\omega(z) = e^{i\alpha}z$, and

(c) $f(|z|<r) \subset F(|z|<r)$ for all r, $0<r<1$.

Proof. (a) $\omega = F^{-1} \circ f$ satisfies the hypotheses of Schwarz's lemma.

(b) $|f'(0)| = |F'(0)||\omega'(0)| \le |F'(0)|$ with equality iff $\omega(z) = e^{i\alpha}z$.

(c) $f(|z|<r) = F(\omega(|z|<r)) \subset F(|z|<r)$.

An immediate application is the following distortion theorem:

THEOREM 1.2. If $f \in P$, then $(1-|z|)/(1+|z|) \le |f(z)| \le (1+|z|)/(1-|z|)$ and $|f'(0)| \le 2$. Equality occurs iff $f(z) = (1+\eta z)/(1-\eta z)$ for some η, $|\eta| = 1$.

Proof. $f \prec (1+z)/(1-z)$.

As a consequence, P is a locally uniformly bounded, hence normal and compact, subset of $H(U)$. Now a coefficient theorem:

THEOREM 1.3. Let $f(z) = 1 + \sum_{n=1}^{\infty} c_n z^n \in P$. Then $|c_n| \le 2$ for all n.

Proof. The function

$$g(z) = \frac{1}{n}\sum_{k=1}^{n} f(e^{2\pi ik/n}z) = 1 + \sum_{j=1}^{\infty} c_j z^j \frac{1}{n}\sum_{k=1}^{n} e^{2\pi ikj/n} = 1 + \sum_{\nu=1}^{\infty} c_{\nu n} z^{\nu n}$$

belongs to P, as well as $G(z) = g(z^{1/n}) = 1 + c_n z + \dots$. Apply Theorem 1.2 to G.

<u>LEMMA 1.4.</u> Let $f(z) = 1 + \sum_{k=1}^{\infty} c_k z^k \in P$, and for fixed $n > 0$ define

$$u(z) = \frac{1}{4i} \sum_{k=0}^{\infty} (c_{-n}c_{k+n} - c_n c_{k-n}) z^k$$

where $c_0 = 2$ and $c_{-k} = \bar{c}_k$. Then $f \pm u \in P$.

<u>Proof.</u> Let $p(z) = 1 + \sum_{k=1}^{n} c_k z^k$, and for $0 < r < 1$ define

$$4i\, u(z,r) = \bar{c}_n z^{-n}[f(rz) - p(rz)] - c_n z^n[f(rz) + \overline{p(r/\bar{z})}] + |c_n|^2.$$

Then

$$\lim_{r \to 1} 4i\, u(z,r) = c_{-n} z^{-n} \sum_{k=n}^{\infty} c_k z^k - c_n z^n \sum_{k=-n}^{\infty} c_k z^k = 4i\, u(z)$$

and

$$4i[f(rz) \pm u(z,r)] = [4i \pm (\bar{c}_n z^{-n} - c_n z^n)] f(rz) \mp [\bar{c}_n z^{-n} p(rz) + c_n z^n \overline{p(r/\bar{z})}] \pm |c_n|^2.$$

For $|z| = 1$, $\mathcal{R}e[f(rz) \pm u(z,r)] = [1 \mp \tfrac{1}{2} \mathcal{I}m\, c_n z^n] \mathcal{R}e\, f(rz) \geq 0$. Therefore $\mathcal{R}e[f(rz) \pm u(z,r)] \geq 0$ for $z \in U$ by the minimum principle. Let $r \to 1$; then $f \pm u \in P$.

<u>DEFINITION.</u> f is an <u>extreme point</u> of a set $A \subset H(D)$ if $f \neq \lambda f_1 + (1-\lambda) f_2$ for all $f_1, f_2 \in A$ $(f_1 \neq f_2)$ and $0 < \lambda < 1$. Denote the set of extreme points of A by E_A.

<u>THEOREM 1.5.</u> $E_P = \{(1 + \eta z)/(1 - \eta z) : |\eta| = 1\}$.

<u>Proof</u>([H8]). Since $1 \pm z \in P$ and $1 = \tfrac{1}{2}(1+z) + \tfrac{1}{2}(1-z)$, it is clear that $1 \notin E_P$. On the other hand, $E_P \neq \emptyset$ by the Krein-Milman theorem (see Appendix A). Therefore let $f \in E_P$ and $c_n = |c_n|\eta$, $|\eta| = 1$, be its first nonzero coefficient. Since $f = \tfrac{1}{2}(f+u) + \tfrac{1}{2}(f-u)$, we must have $u \equiv 0$ in Lemma 1.4. Therefore $c_{k+n} = \eta^2 c_{k-n}$ for all $k \geq 0$. So

$$c_{mn} = \begin{cases} 2\eta^m & \text{if } m \text{ is even} \\ |c_n|\eta^m & \text{if } m \text{ is odd} \end{cases}$$

and all other c_k are zero. That is,

$$f(z) = \tfrac{1}{2}|c_n|(1+\eta z^n)/(1-\eta z^n) + (1-\tfrac{1}{2}|c_n|)[1+(\eta z^n)^2]/[1-(\eta z^n)^2].$$

Since $f \in E_P$, necessarily $|c_n| = 2$ and

$$f(z) = (1+\eta z^n)/(1-\eta z^n) = n^{-1}\sum_{k=1}^{\infty}(1+\eta^{1/n}e^{2\pi ik/n}z)/(1-\eta^{1/n}e^{2\pi ik/n}z)$$

so that $n=1$ also. Thus $E_P \subset \{(1+\eta z)/(1-\eta z) : |\eta| = 1\}$. In fact, the sets are the same since if $(1+\eta_o z)/(1-\eta_o z) = \lambda f_1(z) + (1-\lambda) f_2(z)$, then every $(1+\eta z)/(1-\eta z) = \lambda f_1(\eta\bar{\eta}_o z) + (1-\lambda) f_2(\eta\bar{\eta}_o z)$ and $E_P = \phi$.

REMARK. From the obvious correspondence between P and the class $\mathrm{Re}\, P$ of all nonnegative harmonic functions u in U with $u(0) = 1$, it follows that the extreme points of $\mathrm{Re}\, P$ are $\mathrm{Re}\frac{1+\eta z}{1-\eta z}$ for $|\eta| = 1$.

THEOREM 1.6 (Herglotz representation). If $f \in P$, then there exists a unique (nonnegative) Borel measure μ such that

$$f(z) = \int_{|\eta|=1} (1+\eta z)/(1-\eta z)\, d\mu, \qquad \int_{|\eta|=1} d\mu = 1.$$

Equivalently,

THEOREM 1.6′. If u is harmonic in U, $u > 0$, and $u(0) = 1$, then there exists a unique (nonnegative) Borel measure μ such that

$$u(z) = \int_{|\eta|=1} \mathrm{Re}[(1+\eta z)/(1-\eta z)]\, d\mu, \qquad \int_{|\eta|=1} d\mu = 1.$$

Proof. Apply Choquet's theorem in Appendix A. $E_{\mathrm{Re}P}$ is parametrized in a smooth way by the circle $|\eta| = 1$ so we can transfer the measure there.

APPLICATION. An example of a continuous linear functional $L: H(U) \to C$ is $L(f) = f^{(n)}(z)$ for fixed $z \in U$ and fixed $n \geq 0$. To solve linear extremal problems over P, we only need to examine the extreme points $(1+\eta z)/(1-\eta z)$, $|\eta| = 1$ (see Theorem A.3 in Appendix A). As an application, therefore,

$$|f^{(n)}(z)| \leq 2(n!)/(1-|z|)^{n+1} \quad \text{for all } f \in P,\ z \in U,\ n \geq 0.$$

CHAPTER 2. Special classes: convex, starlike, real, typically real, close-to-convex, bounded boundary rotation

We shall examine some special families for which we can determine the closed convex hulls and corresponding extreme points. In principle then, all linear extremal problems for these families are elementary.

We shall restrict our attention to normalized analytic functions

$$N = \{f \in H(U) : f(0) = 0,\ f'(0) = 1\} .$$

Then
$$\begin{aligned}
S &= H_u(U) \cap N \\
K &= \{f \in S : f(U) \text{ is convex}\} \\
S^* &= \{f \in S : f(U) \text{ is starlike with respect to the origin}\} \\
S_{\mathbb{R}} &= \{f \in S : f^{(n)}(0) \in \mathbb{R} \text{ for all } n \geq 2\} \\
T_{\mathbb{R}} &= \{f \in N : (\mathfrak{Im}\, z)(\mathfrak{Im}\, f(z)) \geq 0\} \\
C &= \{f \in N : \mathfrak{Re}[f'/e^{i\alpha}\varphi'] > 0 \text{ for some } \varphi \in K \text{ and } \alpha \in \mathbb{R}\}
\end{aligned}$$

are the familiar normalized <u>schlicht</u>, <u>convex</u>, <u>starlike</u>, <u>real</u>, <u>typically</u> <u>real</u>, and <u>close-to-convex</u> <u>classes</u>, respectively.

Functions in K and S^* have obvious geometric properties. Functions in $T_{\mathbb{R}}$ have real Maclaurin coefficients. Consequently, if $f \in T_{\mathbb{R}}$ then $f(\bar{z}) = \overline{f(z)}$ and $f(U)$ is symmetric with respect to the real axis. Clearly $S_{\mathbb{R}} \subset T_{\mathbb{R}}$, but $T_{\mathbb{R}}$ contains also some nonunivalent functions (e.g., $z + z^3$) On the other hand, close-to-convex functions turn out to be univalent, and this is the motivation for the definition of C .

<u>LEMMA 2.1</u>. If $F \in H(D)$ where D is convex and $\mathfrak{Re}\, F' > 0$, then $F \in H_u(D)$.

<u>Proof</u>. Suppose $F(z_1) = F(z_2)$ for some $z_1, z_2 \in D$, $z_1 \neq z_2$. Since D is convex, $(1-t)z_1 + tz_2 \in D$ for $t \in [0,1]$. Therefore

$$0 = \Re e \frac{F(z_2)-F(z_1)}{z_2-z_1} = \Re e \frac{1}{z_2-z_1} \int_{z_1}^{z_2} F'(z)dz = \int_0^1 (\Re e F')dt > 0$$

presents a contradiction.

THEOREM 2.2. If $f \in C$, then f is univalent.

Proof. Assume $f \in H(U)$ and $\Re e f'/\Phi' > 0$ where $\Phi = e^{i\alpha}\varphi$, $\varphi \in K$, and $\alpha \in \mathbb{R}$. Then $F = f\circ\Phi^{-1} \in H(D)$ where $D = \Phi(U)$ is convex. Since $\Re e F' = \Re e f'/\Phi' > 0$, by Lemma 2.1 F and $F\circ\Phi = f$ are univalent.

We shall need to convert the geometric definitions for K and S^* into analytic relations.

THEOREM 2.3. If $f \in S$, then the following are equivalent:

(a) $f \in S^*$;

(b) $D_r = f(|z| < r)$ is starlike with respect to the origin for $0 < r < 1$;

(c) $zf'/f \in P$.

Proof. (b) ⇔ (c): D_r is starlike iff $\arg f(re^{i\theta})$ is a non-decreasing function of θ, i.e.,

$$0 \le \frac{\partial}{\partial\theta} \arg f(re^{i\theta}) = \Im m \frac{\partial}{\partial\theta} \log f = \Re e\, zf'/f \;.$$

$\Re e\, zf'/f > 0$ by the minimum principle, since $zf'/f|_{z=0} = 1$.

(b) ⇒ (a) is obvious since each point of $f(U)$ belongs to D_r for some r sufficiently close to 1.

(a) ⇒ (b): For $0 < t < 1$, $\omega(z) = f^{-1}(tf(z))$ satisfies $|\omega(z)| \le |z|$ by Schwarz's lemma. Suppse $w_o \in D_r$. Then $|z_o| = |f^{-1}(w_o)| < r$ and $|f^{-1}(tw_o)| = |\omega(z_o)| \le |z_o| < r$. Therefore $tw_o \in D_r$ for $0 < t < 1$. So D_r is starlike.

THEOREM 2.4. If $f \in S$, then the following are equivalent:

(a) $f \in K$;

(b) $D_r = f(|z| < r)$ is convex for $0 < r < 1$;

(c) $1 + zf''/f' \in P$;

(d) $zf' \in S^*$.

Proof. (b) $\Leftrightarrow$ (c): D_r is convex iff the tangent angle $\arg \dfrac{\partial f(re^{i\theta})}{\partial \theta}$ is a nondecreasing function of θ, i.e.,

$$0 \leq \frac{\partial}{\partial \theta} \arg \frac{\partial f(re^{i\theta})}{\partial \theta} = \mathcal{I}m \frac{\partial}{\partial \theta} \log \frac{\partial f}{\partial \theta} = \mathcal{R}e\,\{1 + zf''/f'\}.$$

$\mathcal{R}e\,\{1 + zf''/f'\} > 0$ by the minimum principle since $[1 + zf''/f']_{z=0} = 1$.

(b) $\Rightarrow$ (a) is obvious since any two points of $f(U)$ belong to some D_r for r sufficiently close to 1.

(a) $\Rightarrow$ (b): Suppose $z_1 \neq z_2$ where $|z_1| \leq |z_2| < r$. Then $w_\lambda = \lambda f(z_1) + (1-\lambda) f(z_2) \in f(U)$ for $0 < \lambda < 1$ since $f(U)$ is convex; hence, $w_\lambda = f(z_\lambda)$ for some $z_\lambda \in U$. We must show that $w_\lambda \in D_r$. Let $g(z) = \lambda f(zz_1/z_2) + (1-\lambda) f(z)$. Then $g \in H(U)$, $g(0) = 0$, and $g(U) \subset f(U)$. By Theorem 1.1, $g(|z| < r) \subset D_r$; in particular, $w_\lambda = g(z_2) \in D_r$.

(c) $\Leftrightarrow$ (d): f satisfies (c) iff $g = zf'$ satisfies part (c) of Theorem 2.3.

COROLLARY 2.5. $C = \{f \in N\colon\ \mathcal{R}e[zf'/e^{i\alpha} g] > 0$ for some $g \in S^*$ and $\alpha \in \mathbb{R}\}$.

Proof. Apply Theorem 2.4(d) to the definition of C.

COROLLARY 2.6. $K \subset S^* \subset C$.

Proof. $K \subset S^*$ is geometrically obvious. $S^* \subset C$ follows by choosing $\alpha = 0$ and $g = f$ in Corollary 2.5.

It will be useful later to know that the analytic conditions of Theorems 2.3(c) and 2.4(c) actually imply univalence. We therefore sketch the proof:

THEOREM 2.7. Let $f \in N$.

(a) If $zf'/f \in P$, then $f \in S^*$.

(b) If $1 + zf''/f' \in P$, then $f \in K$.

Proof. $\gamma_r = f(|z| = r)$ is a closed curve and for $w_o \notin \gamma_r$

$$n(r,w_o) = \frac{1}{2\pi i}\int_{\gamma_r} \frac{dw}{w-w_o} = \frac{1}{2\pi i}\int_{|z|=r} \frac{f'(z)}{f(z)-w_o}dz$$

gives the number of times γ_r winds around w_o and also the number of times f assumes w_o in $|z| < r$.

(a) If $zf'/f \in P$ and $z = re^{i\theta}$, then

$$n(r,0) = \frac{1}{2\pi}\int_0^{2\pi} \frac{zf'}{f}d\theta = \frac{1}{2\pi}\int_0^{2\pi}(1 + \text{positive powers of } z)d\theta = 1$$

so that γ_r winds once around the origin. Since $\frac{\partial}{\partial\theta}\arg f(re^{i\theta}) = \Re e\, zf'/f > 0$, γ_r is a Jordan curve.

(b) If $1 + zf''/f' \in P$ and $z = re^{i\theta}$, then

$$\int_0^{2\pi}\left|\frac{\partial}{\partial\theta}\arg\frac{\partial f}{\partial\theta}\right|d\theta = \int_0^{2\pi}\Re e\left(1 + \frac{zf''}{f'}\right)d\theta = \Re e\int_0^{2\pi}(1 + \text{positive powers of } z)d\theta = 2\pi .$$

So the total variation of the tangent angle to γ_r is 2π. Therefore γ_r can wind around any point at most once and is a Jordan curve. Since the total change in the tangent angle is also 2π, the winding is in the positive direction.

In both cases (a) and (b) the winding numbers $n(r,w_o)$ are identically one for w_o in the interior of γ_r and zero for w_o in the exterior of γ_r. Therefore f assumes each point in the interior of γ_r exactly once in $|z| < r$ and does not assume any point of the exterior of γ_r in $|z| < r$. Since f is an open mapping, f maps $|z| < r$ in a one-to-one fashion onto the interior of γ_r. This is true for all $r < 1$. Therefore f is univalent in U. Since now $f \in S$, the respective assertions follow from Theorems 2.3 and 2.4 .

THEOREM 2.8 (Ruscheweyh and Sheil-Small [R47]). If $f \in K$, then

$$\mathrm{Re}\left\{\frac{z}{z-\zeta}\,\frac{\zeta-t}{z-t}\,\frac{f(z)-f(t)}{f(\zeta)-f(t)} - \frac{\zeta}{z-\zeta}\right\} > \frac{1}{2} \quad \text{for all } z,\zeta,t \in U .$$

Proof. The function $F(z,\zeta,t) = \frac{2z}{z-\zeta}\,\frac{\zeta-t}{z-t}\,\frac{f(z)-f(t)}{f(\zeta)-f(t)} - \frac{2\zeta}{z-\zeta} - 1$ is analytic in all three variables when extended by continuity for $z=\zeta$, $z=t$, and $\zeta=t$. If $0<\alpha,\beta<2\pi$, $\alpha\neq\beta$, then

$$\mathrm{Re}\, F(e^{i\alpha}t,\, e^{i\beta}t,\, t) = \frac{1}{\sin\frac{1}{2}(\alpha-\beta)}\,\frac{\sin\frac{1}{2}\beta}{\sin\frac{1}{2}\alpha}\,\mathrm{Im}\left\{\frac{f(e^{i\alpha}t)-f(t)}{f(e^{i\beta}t)-f(t)}\right\} .$$

Since $f(|z|\le r)$ is convex for $r=|t|$,

$$\arg \frac{f(e^{i\alpha}t)-f(t)}{f(e^{i\beta}t)-f(t)} \in \begin{cases} [0,\pi] & \text{if } 0<\beta<\alpha<2\pi \\ [-\pi,0] & \text{if } 0<\alpha<\beta<2\pi \end{cases} .$$

Therefore $\sin\frac{1}{2}(\alpha-\beta)$ and $\mathrm{Im}\left\{\frac{f(e^{i\alpha}t)-f(t)}{f(e^{i\beta}t)-f(t)}\right\}$ have the same sign so that $\mathrm{Re}\, F(e^{i\alpha}t, e^{i\beta}t,t)\ge 0$. By continuity, $\mathrm{Re}\, F(z,\zeta,t)\ge 0$ whenever $|z|=|\zeta|=|t|$. Since $\mathrm{Re}\, F(z,\zeta,t)$ is a harmonic function of all three variables, the minimum principle implies $\mathrm{Re}\, F\ge 0$ for all $z,\zeta,t\in U$. Actually $\mathrm{Re}\, F>0$ since $F(0,0,0)=1$. (To apply the minimum principle one can observe that $\{|z|=r\}^3$ is the <u>distinguished</u> <u>boundary</u> of the polydisk $\{|z|<r\}^3$, or else one can apply it successively in the variables z,ζ,t individually to obtain $\mathrm{Re}\, F>0$ first for $|z|\le|\zeta|=|t|$, then for $|\zeta|,|z|\le|t|$, and finally for $|t|,|\zeta|,|z|<1$.)

COROLLARY 2.9 (Suffridge [S23]). If $f\in K$, then

$$\mathrm{Re}\left\{\frac{zf'(z)}{f(z)-f(\zeta)} - \frac{\zeta}{z-\zeta}\right\} > \frac{1}{2} \quad \text{for all } z,\zeta\in U .$$

Proof. Let $t\to z$ in Theorem 2.8 .

If we also let $\zeta\to z$, the above inequality implies that

$1+zf''/f' \in P$. It follows then from Theorem 2.7(b) that the conditions of Theorem 2.8 and Corollary 2.9 are not only necessary, but also sufficient for $f \in N$ to belong to K. In fact, the expressions are invariant when replacing f by $Af+B$. They therefore characterize the convexity of $f(U)$ independent of normalization.

COROLLARY 2.10 (Strohäcker [S21], Marx[M1]). If $f \in K$, then

$$\Re e\, zf'/f > \tfrac{1}{2} \quad \text{and} \quad \Re e\, f/z > \tfrac{1}{2}\,.$$

Proof. Set $\zeta = 0$ in Corollary 2.9, and let $t=0$, $\zeta \to 0$ in Theorem 2.8.

f is called starlike of order α if $\Re e\, zf'/f > \alpha$. Therefore convex mappings are starlike of order $\tfrac{1}{2}$. More importantly, the second condition of Corollary 2.10 turns out to describe precisely the closed convex hull of K. This and related results that follow are due to L. Brickman, T. H. MackGregor, and D. R. Wilken [B10].

THEOREM 2.11 ([B10]). K and $\overline{co}\,K$ are compact,

$$\overline{co}\,K = \{f \in N : \Re e\, f/z > \tfrac{1}{2}\}$$

$$= \{\int_{|\eta|=1} z/(1-\eta z)\,d\mu : \mu \text{ is a probability measure on } |\eta|=1\},$$

and $E_{\overline{co}\,K} = \{z/(1-\eta z) : |\eta| = 1\}$.

Proof. The mapping $\mathcal{L}$ defined by $\mathcal{L}(g) = \tfrac{1}{2}z(1+g)$ is a linear homeomorphism of $H(U)$ onto the subspace $\{h \in H(U) : h(0) = 0\}$. By Theorem A.2 it maps the compact convex set P onto a compact convex set $\mathcal{L}(P) = \{f \in N : \Re e\, f/z > \tfrac{1}{2}\}$. A second representation for $\mathcal{L}(P)$ comes from Theorem 1.6; namely, $f \in \mathcal{L}(P)$ iff there is a probability measure μ such that

$$f(z) = \tfrac{1}{2}z \int_{|\eta|=1} [1 + (1+\eta z)/(1-\eta z)]\,d\mu = \int_{|\eta|=1} z/(1-\eta z)\,d\mu\,.$$

It follows from Corollary 2.10 that K is a subset of $\mathcal{L}(P)$. K is closed since the relation of Theorem 2.4(c) is preserved under locally uniform convergence of functions in K. Consequently, K is a compact subset of the compact convex set $\mathcal{L}(P)$. Therefore $\overline{co}\, K \subset \mathcal{L}(P)$.

By Theorems A.2 and 1.5, $E_{\mathcal{L}(P)} = \mathcal{L}(E_P) = \{z/(1-\eta z): \; |\eta| = 1\}$. Observe that the mappings $z/(1-\eta z)$ belong to K. Hence $E_{\mathcal{L}(P)} \subset K$ and $\mathcal{L}(P) = \overline{co}\, E_{\mathcal{L}(P)} \subset \overline{co}\, K$ by the Krein-Milman theorem (Appendix A). Therefore $\mathcal{L}(P) = \overline{co}\, K$, and the proof is complete.

The extreme points $z/(1-\eta z)$ map U onto the half-planes whose boundaries have distance $\frac{1}{2}$ from the origin.

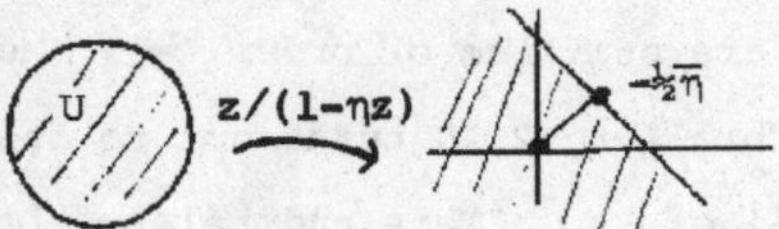

Since the extreme values of the functional $L(f) = f^{(n)}(z)$ must occur at an extreme point (Theorem A.3), we have the following immediate application.

<u>THEOREM 2.12</u>. If $f(z) = z + \sum_{n=2}^{\infty} c_n z^n \in K$, then

$$|f(z)| \le |z|/(1-|z|) \quad \text{and} \quad |f^{(n)}(z)| \le n!/(1-|z|)^{n+1} \quad \text{for all } z \in U \text{ and } n \ge 1 .$$

In particular, $|c_n| \le 1$ for $n = 2,3,\ldots$.

We turn to the class S^*.

<u>THEOREM 2.13</u>. $f \in S^*$ iff there exists a probability measure μ such that

$$f(z) = z \exp[-2 \int_{|\eta|=1} \log(1-\eta z)\, d\mu] .$$

Moreover, the probability measure μ is unique.

<u>Proof</u>. If $f \in S^*$, then by Theorems 2.3(c) and 1.6 there is a unique probability measure μ such that

$$[zf'/f-1]/z = \int_{|\eta|=1} [(1+\eta z)/(1-\eta z)-1]/z \, d\mu = \int_{|\eta|=1} 2\eta/(1-\eta z) \, d\mu \, .$$

Therefore $\log f/z = -2 \int_{|\eta|=1} \log(1-\eta z) \, d\mu$ by integration.

Conversely, if f has the given form, then $zf'/f \in P$ and $f \in S^*$ by Theorem 2.7(a).

<u>COROLLARY 2.14</u>. If $f \in S^*$, then $\lim_{r\to 1} \arg f(re^{i\theta})$ exists for all θ.

<u>Proof</u>. Represent $\arg f/z = -2 \int_{|\eta|=1} \arg(1-\eta z) \, d\mu$. The radial limit exists by the Lebesgue bounded convergence theorem.

To determine the $\overline{\text{co}}\, S^*$ we shall exploit the connection with convex mappings rather than Theorem 2.13.

<u>THEOREM 2.15</u> ([B10]). S^* and $\overline{\text{co}}\, S^*$ are compact,

$$\overline{\text{co}}\, S^* = \{ \int_{|\eta|=1} z/(1-\eta z)^2 d\mu \; : \; \mu \text{ is a probability measure on } |\eta|=1 \} ,$$

and

$$E_{\overline{\text{co}}\, S^*} = \{ z/(1-\eta z)^2 \; : \; |\eta|=1 \} .$$

<u>Proof</u>. The mapping $\mathcal{L}$ defined by $\mathcal{L}(g) = zg'$ is a linear homeomorphism of the space $\{h \in H(U) : h(0) = 0\}$, and $\mathcal{L}(K) = S^*$ by Theorems 2.4(d) and 2.7. The results now follow from applying Theorem A.2 to Theorem 2.11.

The starlike functions $z/(1-\eta z)^2$ are called <u>Koebe</u> <u>functions</u> and map U onto the complement of a ray from $-\frac{1}{4}\bar{\eta}$ to ∞ :

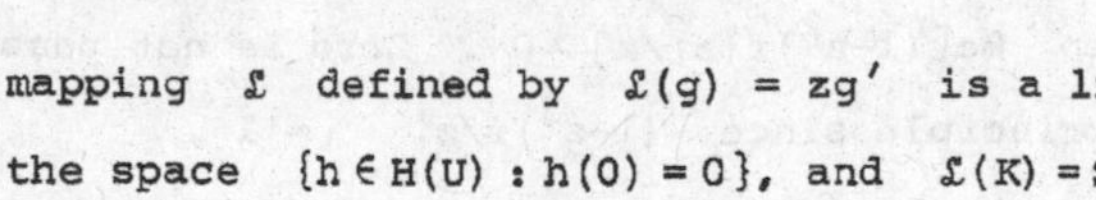

By examining just these extreme points we have the following application.

<u>THEOREM 2.16</u>. If $f(z) = z + \sum_{n=2}^{\infty} a_n z^n \in S^*$, then

$$|f^{(n)}(z)| \le n!(n+|z|)/(1-|z|)^{n+2} \quad \text{for all } z \in U \text{ and } n \ge 0 .$$

In particular, $|a_n| \le n$ for $n = 2,3,\ldots$.

We turn now to the classes $S_{\mathbb{R}}$ and $T_{\mathbb{R}}$. We shall see that $T_{\mathbb{R}}$ is precisely the closed convex hull of $S_{\mathbb{R}}$.

<u>THEOREM 2.17</u>. The following are equivalent:

(a) $f \in T_{\mathbb{R}}$;

(b) $(1-z^2)f/z \in P$ and $f^{(n)}(0) \in \mathbb{R}$ for all $n \ge 2$;

(c) there exists a (unique) probability measure μ on $[-1,1]$ such that

$$f(z) = \int_{[-1,1]} z/(1-2xz+z^2)\,d\mu .$$

<u>Proof</u>. (a) $\Rightarrow$ (b): If $f \in T_{\mathbb{R}}$, then by continuity f is real if z is real so that $f^{(n)}(0) \in \mathbb{R}$ for all n. If $0 < r < 1$ and $|z| = 1$, then

$$\Re e[(1-z^2)rf(rz)/z] = 2\,\Im m(rz) \cdot \Im m\, f(rz) \ge 0 .$$

By the minimum principle, $\Re e[(1-z^2)rf(rz)/z] \ge 0$ in U. Let $r \to 1$; then $\Re e[(1-z^2)f(z)/z] \ge 0$. Zero is not possible by the minimum principle since $(1-z^2)f/z|_{z=0} = 1$.

(b) $\Rightarrow$ (c): Since all Maclaurin coefficients of f are real,

$$f(z) = \tfrac{1}{2}[f(z) + \overline{f(\bar z)}] = \tfrac{1}{2}z(1-z^2)^{-1}[p(z) + \overline{p(\bar z)}]$$

where $p \in P$. By Theorem 1.6 there is a probability measure ν such that

$$f(z) = \int_{|\eta|=1} z/(1-2(\Re e\,\eta)z+z^2)\,d\nu = \int_{[-1,1]} z/(1-2xz+z^2)\,d\mu$$

where "$\mu(x) = \nu(\eta) + \nu(\bar\eta)$," i.e., $\mu(A) = \nu\{e^{i\theta}: \cos\theta \in A\}$ for $A \subset [-1,1]$. For the uniqueness we assume $f(z) = \int_{[-1,1]} z/(1-2xz+z^2)\,d\mu_k$, k=1,2, and define "$\nu_k(\eta) = \frac{1}{2}\mu_k(\mathrm{Re}\,\eta)$" so that "$\nu_k(\eta) = \nu_k(\bar{\eta})$," i.e.,

$$\nu_k(B) = \tfrac{1}{2}\mu_k(\{\cos\theta : e^{i\theta} \in B,\ 0 \le \theta \le \pi\}) + \tfrac{1}{2}\mu_k(\{\cos\theta : e^{-i\theta} \in B,\ 0 \le \theta \le \pi\})$$

for $B \subset \{|\eta| = 1\}$. Then $\int_{|\eta|=1} (1+\eta z)/(1-\eta z)\, d\nu_k = (1-z^2) f/z$, and by the uniqueness of the Herglotz representation $\nu_1 = \nu_2$; hence $\mu_1 = \mu_2$.

(c) ⇒ (a): It is clear that $f \in N$. Since $\mathrm{Im}[z/(1-2xz+z^2)] = (1-|z|^2)|1-2xz+z^2|^{-2}\, \mathrm{Im}\, z$ has the same sign as $\mathrm{Im}\, z$, it follows that $f \in T_{\mathbb{R}}$.

<u>THEOREM 2.18</u> ([B10]). $S_{\mathbb{R}}$ and $T_{\mathbb{R}}$ are compact, $T_{\mathbb{R}}$ is also convex, $\overline{\mathrm{co}}\, S_{\mathbb{R}} = T_{\mathbb{R}}$, and

$$E_{T_{\mathbb{R}}} = \{z/(1-2xz+z^2) : x \in [-1,1]\}.$$

<u>Proof</u>. That $S_{\mathbb{R}}$ and $T_{\mathbb{R}}$ are closed, $T_{\mathbb{R}}$ is convex, and $S_{\mathbb{R}} \subset T_{\mathbb{R}}$ is clear from their definitions. By Theorem 2.17(b), $T_{\mathbb{R}} \subset \{zp/(1-z^2) : p \in P\}$, which is compact since P is compact. Hence both $S_{\mathbb{R}}$ and $T_{\mathbb{R}}$ are compact and $\overline{\mathrm{co}}\, S_{\mathbb{R}} \subset T_{\mathbb{R}}$.

Since the representation in Theorem 2.17(c) is unique, $E_{T_{\mathbb{R}}} = \{z/(1-2xz+z^2) : x \in [-1,1]\}$. In fact, each function $z/(1-2xz+z^2)$ belongs to $S_{\mathbb{R}}$ so that by the Krein-Milman theorem $T_{\mathbb{R}} = \overline{\mathrm{co}}\, E_{T_{\mathbb{R}}} \subset \overline{\mathrm{co}}\, S_{\mathbb{R}}$; therefore $\overline{\mathrm{co}}\, S_{\mathbb{R}} = T_{\mathbb{R}}$.

The functions $z/(1-2xz+z^2)$ map U onto the complement of real slits from $-\frac{1}{2}(1+x)^{-1}$ to $-\infty$ and from $\frac{1}{2}(1-x)^{-1}$ to ∞. (Note for $x = \pm 1$ one of the slits disappears and we have Koebe functions.)

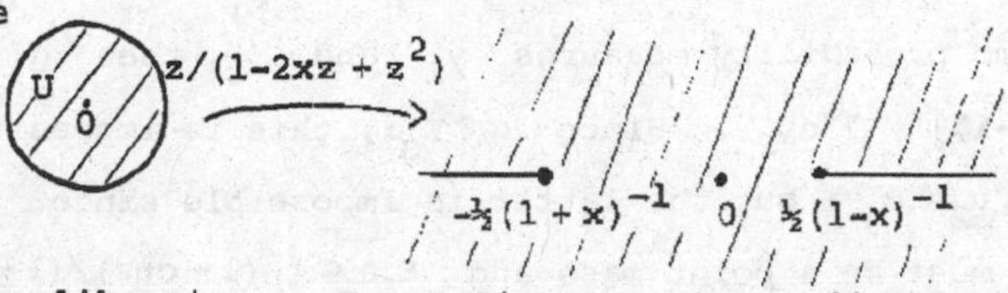

These mappings are starlike, i.e., $E_{T_{\mathbb{R}}} \subset S^*$, so as an application we have the estimates of Theorem 2.16:

<u>THEOREM 2.19</u>. If $f(z) = z + \sum_{n=2}^{\infty} a_n z^n \in T_{\mathbb{R}}$, then

$|f^{(n)}(z)| \le n!(n+|z|)/(1-|z|)^{n+2}$ for all $z \in U$ and $n \ge 0$.
In particular, $|a_n| \le n$ for $n = 2,3,\ldots$.

The following theorem is a fundamental tool in the further study of special families.

THEOREM 2.20 (Brannan, Clunie, and Kirwan [B8]). Let $F = \{f \in H(U) : f \prec (1+cz)/(1-z)\}$ for fixed c, $|c| \le 1$. For $\alpha \ge 1$ let $F^\alpha = \{f^\alpha : f \in F\}$ and $G^\alpha = \{\int_{|\eta|=1} [(1+c\eta z)/(1-\eta z)]^\alpha d\mu : \mu$ is a probability measure on $|\eta| = 1\}$. Then F, F^α, and G^α are compact, $\overline{co}\, F^\alpha = G^\alpha$, and $E_{G^\alpha} = \{[(1+c\eta z)/(1-\eta z)]^\alpha : |\eta|=1\}$.

Proof. Since $(1+cz)/(1-z) = \frac{1}{2}(1+c)(1+z)/(1-z) + \frac{1}{2}(1-c)$, F is an affine image of P. Therefore F is compact and $E_F = \{(1+c\eta z)/(1-\eta z) : |\eta| = 1\}$. Since $|c| \le 1$, the functions in F do not vanish so that F^α is well defined and compact. The convex set G^α is compact since the set of probability measures on $|\eta| = 1$ is weakly compact.

Suppose now $g \in E_{G^\alpha}$. Let $\mathcal{P}_g$ be the set of probability measures μ on $|\eta| = 1$ such that $\int_{|\eta|=1} [(1+c\eta z)/(1-\eta z)]^\alpha d\mu = g(z)$. $\mathcal{P}_g \ne \emptyset$ and is compact in the weak topology. Therefore $E_{\mathcal{P}_g} \ne \emptyset$ by the Krein-Milman theorem. Let $\nu \in E_{\mathcal{P}_g}$. If $\nu = \lambda\nu_1 + (1-\lambda)\nu_2$ for some probability measures ν_1 and ν_2, then $g(z) = \lambda\int[\quad]^\alpha d\nu_1 + (1-\lambda)\int[\quad]^\alpha d\nu_2$. Since $g \in E_{G^\alpha}$, this is impossible unless $\nu_1, \nu_2 \in \mathcal{P}_g$; but the latter is impossible since $\nu \in E_{\mathcal{P}_g}$. Therefore ν must be a point mass and $E_{G^\alpha} \subset \{[(1+c\eta z)/(1-\eta z)]^\alpha : |\eta|=1\}$. The sets are actually equal since if $[(1+c\eta_0 z)/(1-\eta_0 z)]^\alpha = \lambda g_1(z) + (1-\lambda)g_2(z)$, then every $[(1+c\eta z)/(1-\eta z)]^\alpha = \lambda g_1(\eta\bar{\eta}_0 z) + (1-\lambda)g_2(\eta\bar{\eta}_0 z)$ and $E_{G^\alpha} = \emptyset$.

If $f \in F-E_F$, then distinct $f_1, f_2 \in F$ and $\lambda \in (0,1)$ exist such that $f = \lambda f_1 + (1-\lambda) f_2$. Now $f^\alpha = \lambda f^{\alpha-1} f_1 + (1-\lambda) f^{\alpha-1} f_2$ where $f^{\alpha-1} f_k$, $k = 1,2$, are distinct. Since $\log[(1+cz)/(1-z)]$ maps U onto a convex set and $\alpha \geq 1$,

$$(1-1/\alpha)\log f + (1/\alpha)\log f_k < \log[(1+cz)/(1-z)]$$

so that $f^{1-1/\alpha} f_k^{1/\alpha} \in F$ and $f^{\alpha-1} f_k \in F^\alpha$. Therefore $f^\alpha \notin E_{F^\alpha}$. Consequently, $E_{F^\alpha} \subset \{f^\alpha : f \in E_F\} = E_{G^\alpha} \subset F^\alpha$. By the Krein-Milman theorem, $\overline{co}\, F^\alpha \subset G^\alpha \subset \overline{co}\, F^\alpha$.

The next theorem was verified in many cases in [B8] and in general first by D. Aharonov and S. Friedland [A3]. The following elementary proof is due to D. A. Brannan [B7].

<u>THEOREM 2.21</u>. Suppose $f < (1+cz)/(1-z)$ for some c, $|c| \leq 1$. Then for $\alpha \geq 1$ the coefficients of f^α are dominated by the corresponding coefficients of $[(1+z)/(1-z)]^\alpha$.

<u>Proof</u>. By $g \underset{n}{\ll} h$ we mean that the coefficients of z^k for g are dominated by the corresponding coefficients of h for $0 \leq k \leq n$. Fix $\alpha \geq 1$. If $|c| \leq 1$ and $f < (1+cz)/(1-z)$, then $f^\alpha \underset{0}{\ll} [(1+z)/(1-z)]^\alpha$ is obvious. We assume $f^\alpha \underset{n}{\ll} [(1+z)/(1-z)]^\alpha$ and employ induction. If $|c| \leq 1$, then

$$[(1+cz)/(1-z)]^{1-1/\alpha}[(1+z)/(1-z)]^{1/\alpha} < (1+\gamma z)/(1-z)$$

for some γ, $|\gamma| \leq 1$. Therefore

$$[(1+cz)/(1-z)]^{\alpha-1}[(1+z)/(1-z)] \underset{n}{\ll} [(1+z)/(1-z)]^\alpha$$

by the induction hypothesis. Since $[(1+z)/(1-z)]^\alpha$ and $1/(1-z^2)$ have all nonegative coefficients, it follows that

$$\frac{(1+cz)^{\alpha-1}}{(1-z)^{\alpha+1}} = \left(\frac{1+cz}{1-z}\right)^{\alpha-1}\left(\frac{1+z}{1-z}\right)\frac{1}{(1-z^2)} \underset{n}{\ll} \left(\frac{1+z}{1-z}\right)^{\alpha}\frac{1}{(1-z^2)} = \frac{(1+z)^{\alpha-1}}{(1-z)^{\alpha+1}}.$$

Now

$$\begin{aligned}\frac{\partial}{\partial z}[(1+cz)/(1-z)]^\alpha &= \alpha(1+c)(1+cz)^{\alpha-1}/(1-z)^{\alpha+1} \\ &\underset{n}{\ll} 2\alpha(1+z)^{\alpha-1}/(1-z)^{\alpha+1}.\end{aligned}$$

By integration, $[(1+cz)/(1-z)]^{\alpha} \underset{n+1}{<<} [(1+z)/(1-z)]^{\alpha}$. If now $f < (1+cz)/(1-z)$, then by Theorems A3 and 2.20 we may dominate the coefficients of f^{α} by dominating the coefficients of the extreme points $[(1+c\eta z)/(1-\eta z)]^{\alpha}$, $|\eta| = 1$. Since $[(1+c\eta z)/(1-\eta z)]^{\alpha} \underset{n+1}{<<} [(1+z)/(1-z)]^{\alpha}$, we have $f^{\alpha} \underset{n+1}{<<} [(1+z)/(1-z)]^{\alpha}$, and the induction is complete.

Recall that f is close-to-convex if there exist $\varphi \in K$ and $\alpha \in \mathbb{R}$ such that $\mathrm{Re}[f'/e^{i\alpha}\varphi'] > 0$, or equivalently, $|\arg[f'/e^{i\alpha}\varphi']| < \frac{1}{2}\pi$. We first extend this definition in a way that will be useful also for functions of bounded boundary variation.

<u>DEFINITION</u>. The class of normalized <u>close-to-convex</u> <u>functions</u> <u>of</u> <u>order</u> <u>β</u> $(\beta > 0)$ is

$$C(\beta) = \{f \in N: \ |\arg[f'/e^{i\alpha}\varphi']| < \tfrac{1}{2}\beta\pi \ \text{ for some } \varphi \in K \text{ and } \alpha \in \mathbb{R}\}.$$

Note that $C(1)$ is the old close-to-convex class C. Also $f \in C(\beta)$ iff $f \in N$ and $f' = e^{i\alpha}p^{\beta}\varphi'$ for some $\alpha \in \mathbb{R}$, $\varphi \in K$, and $p \in H(U)$ with $\mathrm{Re}\, p > 0$.

Denote

$$k(z;\xi,\eta,\beta) = \begin{cases} \{[(1-\xi z)/(1-\eta z)]^{\beta+1} - 1\}/\{(\beta+1)(\eta-\xi)\} & \text{if } \xi \neq \eta \\ z/(1-\eta z) & \text{if } \xi = \eta \end{cases}$$

where $|\xi| = |\eta| = 1$. Then

$$\frac{d}{dz}k(z;\xi,\eta,\beta) = (1-\xi z)^{\beta}/(1-\eta z)^{\beta+2}$$

Observe that $k(z;\xi,\eta,\beta) \in \dot{C}(\beta)$, for choose $\varphi(z) = z/(1-\eta z) \in K$ and $\alpha \in \mathbb{R}$ such that $\mathrm{Re}[e^{-i\alpha/\beta}(1-\xi z)/(1-\eta z)] > 0$ in U.

THEOREM 2.22([B8]). Let $\beta \ge 1$ and $T = \{|\xi| = 1\} \times \{|\eta| = 1\}$. Then $C(\beta)$ and $\overline{co}C(\beta)$ are compact,

$$\overline{co}C(\beta) = \{\int_T k(z;\xi,\eta,\beta)d\mu : \mu \text{ is a probability measure on } T\},$$

and $E_{\overline{co}C(\beta)} \subset \{k(z;\xi,\eta,\beta) : |\xi| = |\eta| = 1, \ \xi \ne \eta\}$.

Proof. If $f \in C(\beta)$, then $f' = e^{i\alpha}p^{\beta}\varphi'$ where $\varphi \in K$ and $\Re e\, p > 0$. Since $f \in N$, $1 = e^{i\alpha}p(0)^{\beta}$ and $e^{i\alpha/\beta}p \prec (1+cz)/(1-z)$ for an appropriate choice of c with $|c| = 1$. Since $\beta \ge 1$, there is by Theorem 2.20 a probability measure μ_1 on $|\xi| = 1$ such that

$$e^{i\alpha}p(z)^{\beta} = \int_{|\xi|=1}[1+c\xi z)/(1-\xi z)]^{\beta}d\mu_1 .$$

By Theorem 2.11 there is a probability measure μ_2 on $|\eta| = 1$ such that $\varphi'(z) = \int_{|\eta|=1} 1/(1-\eta z)^2 d\mu_2$. Therefore

$$f'(z) = \int_T [(1+c\xi z)/(1-\xi z)]^{\beta}/(1-\eta z)^2 \, d\mu_1 \times \mu_2 .$$

Since $\log(1-z)$ maps U onto a convex set,

$$[\beta/(\beta+2)]\log(1-\xi z) + [2/(\beta+2)]\log(1-\eta z) \prec \log(1-z)$$

and $1/[1-\xi z)^{\beta/(\beta+2)}(1-\eta z)^{2/(\beta+2)}] \prec 1/(1-z)$. By Theorem 2.20 (with $c = 0$ and $\alpha = \beta+2$) there is a probability measure μ_3 on $|\zeta| = 1$ such that

$$1/[(1-\xi z)^{\beta}(1-\eta z)^2] = \int_{|\zeta|=1} 1/(1-\zeta z)^{\beta+2}d\mu_3 .$$

Therefore

$$[(1+c\xi z)/(1-\xi z)]^{\beta}/(1-\eta z)^2 = \int_{|\zeta|=1}(1+c\xi z)^{\beta}/(1-\zeta z)^{\beta+2}d\mu_3$$

belongs to $B' = \{\int_T \frac{d}{dz}k(z;\xi,\eta,\beta)d\mu : \int_T d\mu = 1\}$. Since B' is convex and compact (the set of probability measures on T is compact in the weak topology), $f' \in B'$. By integration, $f \in B = \{\int_T k(z;\xi,\eta,\beta) : \int_T d\mu = 1\}$.

Clearly B is also convex and compact. Hence $\overline{co}C(\beta) \subset B$.

$C(\beta)$ is a closed subset of B; hence compact. (Indeed, if $f_n \in C(\beta)$, $|\arg[f_n'/e^{i\alpha_n}\varphi_n']| < \frac{1}{2}\beta\pi$, $\varphi_n \in K$, and $f_n \to f$, then for suitable subsequences $e^{i\alpha_{n_k}} \to e^{i\alpha}$, $\varphi_{n_k} \to \varphi \in K$, and $|\arg[f'/e^{i\alpha}\varphi']| \le \frac{1}{2}\beta\pi$. If the latter is a strict inequality, then $f \in C(\beta)$; if not, then by the maximum or minimum principle $f = \varphi \in C(\beta)$.)

Just as in the proof of Theorem 2.20, extreme points of B must be generated by point masses. Therefore $E_B \subset \{k(z;\xi,\eta,\beta) : |\xi| = |\eta| = 1\} \subset C(\beta)$. By the Krein-Milman theorem, $B \subset \overline{co}C(\beta)$. Hence $\overline{co}C(\beta) = B$.

Now $E_{\overline{co}C(\beta)} \subset \{k(z;\xi,\eta,\beta): |\xi| = |\eta| = 1\}$. However, for $\xi = \eta$ the function $k(z;\eta,\eta,\beta) = z/(1-\eta z)$ cannot be an extreme point of $\overline{co}C(\beta)$ since it is not even an extreme point of the subset $\overline{co}S^*$. (Observe the chain $K \subset S^* \subset C \subset C(\beta)$ for $\beta \ge 1$.)

PROBLEM. Determine $\overline{co}C(\beta)$ and $E_{\overline{co}C(\beta)}$ for $0 < \beta < 1$.

PROBLEM. Show that each $k(z;\xi,\eta,\beta) \in E_{\overline{co}C(\beta)}$ for $\xi \ne \eta$, $|\xi| = |\eta| = 1$, $\beta > 1$. For $\beta = 1$ this fact is contained in the following theorem on the close-to-convex class.

THEOREM 2.23 ([B10]). C and $\overline{co}C$ are compact,

$$\overline{co}C = \{\int_T [z-\tfrac{1}{2}(\xi+\eta)z^2]/(1-\eta z)^2 d\mu : \mu \text{ is a probability measure on } T\}$$

where $T = \{|\xi| = 1\} \times \{|\eta| = 1\}$, and

$$E_{\overline{co}C} = \{[z-\tfrac{1}{2}(\xi+\eta)z^2]/(1-\eta z)^2 : |\xi| = |\eta| = 1,\ \xi \ne \eta\} .$$

<u>Proof</u>. Since $C = C(1)$ and $k(z;\xi,\eta,1) = [z-\frac{1}{2}(\xi+\eta)z^2]/(1-\eta z)^2$, everything is a special case of Theorem 2.22 for $\beta = 1$ except the assertion about extreme points. For $|\xi_o| = |\eta_o| = 1$, $\xi_o \neq \eta_o$, we must show that

$$[z-\tfrac{1}{2}(\xi_o+\eta_o)z^2]/(1-\eta_o z)^2 = \int_T [z-\tfrac{1}{2}(\xi+\eta)z^2]/(1-\eta z)^2 d\mu, \quad \int_T d\mu = 1,$$

is possible only if μ is a point mass at (ξ_o, η_o). Taking derivatives we have $(1-\xi_o z)/(1-\eta_o z)^3 = \int_T (1-\xi z)/(1-\eta z)^3 d\mu$. Set $z = r\bar{\eta}_o$, and let $r \to 1$. Then

$$1-\xi_o\bar{\eta}_o = \lim_{r\to 1} \int_T (1-\xi\bar{\eta}_o r)(1-r)^3/(1-\eta\bar{\eta}_o r)^3 d\mu .$$

By the Lebesgue bounded convergence theorem we have $1-\xi_o\bar{\eta}_o = \int_{T_o} (1-\xi\bar{\eta}_o) d\mu$ where $T_o = \{|\xi| = 1\} \times \{\eta_o\}$. Set $a = \int_{T_o} d\mu$. Then $0 \le a \le 1$ and $\xi_o\bar{\eta}_o - (1-a) = \bar{\eta}_o \int_{T_o} \xi d\mu$. On the one hand, $|\bar{\eta}_o \int_{T_o} \xi d\mu| \le a$, and on the other hand, $|\xi_o\bar{\eta}_o - (1-a)| \ge 1-(1-a) = a$. Therefore both inequalities must be equality. Since $\xi_o\bar{\eta}_o \neq 1$, the latter is possible only if $a = 1$. Moreover, $\xi_o\bar{\eta}_o = \bar{\eta}_o \int_{T_o} \xi d\mu$ is possible only if μ is a unit point mass at (ξ_o, η_o).

By expressing the extreme points of C in the form $z/(1-\eta z) - \frac{1}{2}(\xi-\eta)z^2/(1-\eta z)^2$ and their n^{th} derivatives as $n!\,\eta^{n-2}[\frac{1}{2}(n+1)\eta - \frac{1}{2}(n-1)\xi - \xi\eta z]/(1-\eta z)^{n+2}$ and using elementary estimates, we have as an application the following extension of Theorem 2.16.

<u>THEOREM 2.24</u>. If $f(z) = z + \sum_{n=2}^{\infty} a_n z^n \in C$, then

$$|f^{(n)}(z)| \le n!(n+|z|)/(1-|z|)^{n+2} \quad \text{for all } z \in U \text{ and } n \ge 0 .$$

In particular, $|a_n| \le n$ for $n = 2, 3, \ldots$.

PROBLEMS. There are many intriguing extremal problems that are not linear. One example is the conjecture of M. S. Robertson [R3] that

$$\left| n|a_n| - m|a_m| \right| \le |n^2-m^2|$$

holds for the coefficients of close-to-convex functions. J. A. Jenkins [J2] has shown that the conjecture fails in the full class S. However, Robertson [R2] has verified the conjecture in C in case $n-m$ is an even integer and in some other cases. It is sufficient to find a single functional $\chi : S^* \to C$ such that

$$|a_n - \chi a_{n-1}| \le 1$$

holds in the starlike class S^*, or equivalently, to find some functional $\tilde{\chi} : K \to C$ such that

$$|na_n - (n-1)\tilde{\chi}a_{n-1}| \le 1$$

holds in the convex class K for all n. Perhaps the inequality of Theorem 2.8 can be used profitably to attack the last problem.

EXERCISE. A natural candidate for the functional $\chi : S^* \to C$ is $\frac{1}{2}a_2$ since it works for $E_{\overline{\cos}}{}^*$. However, verify that the inequalities

$$|a_n - \tfrac{1}{2}a_2a_{n-1}| \le 1$$

are violated for each $n \ge 4$ by the functions $z(1+z)^{-\epsilon}(1-z)^{\epsilon-2} \in S^*$ if ϵ is positive, but sufficiently small. This observation is due to R. Barnard.

We will call σ a finite signed measure if $\sigma = \mu - \nu$ where μ and ν are finite (nonnegative regular Borel) measures and denote its corresponding variation measure by $|\sigma|$.

DEFINITION. The class of functions of bounded boundary rotation is

$$V = \{\int_0^z \exp[\int_{|\eta|=1} \log(1-\eta z)\,d\sigma]\,dz : \sigma \text{ is a finite signed measure on } |\eta|=1 \text{ and } \int_{|\eta|=1} d\sigma = -2\} .$$

V_k is the subclass of V for which $\int_{|\eta|=1} d|\sigma| \le k \quad (k \ge 2)$.

By Theorems 2.13 and 2.4(d), V_2 is just the convex class K.

For $f \in V_k$ the constant k restricts the rotation of the boundary in the following sense. The tangent to the level curve $\gamma_r = f(|z|=r)$ at $f(re^{i\theta})$ is $\frac{\partial}{\partial\theta} f(re^{i\theta})$. The variation of the tangent along γ_r is then

$$\int_0^{2\pi} d_\theta |\arg \frac{\partial}{\partial\theta} f(re^{i\theta})| = \int_0^{2\pi} |\frac{\partial}{\partial\theta} \arg \frac{\partial}{\partial\theta} f(re^{i\theta})|\,d\theta = \int_0^{2\pi} |\mathrm{Re}(1 + \frac{zf''}{f'})|\,d\theta .$$

Since $f \in V$, the harmonic function $u = \mathrm{Re}(1 + \frac{zf''}{f'}) = \frac{1}{2}\int_{|\eta|=1} \mathrm{Re}[(1+\eta z)/(1-\eta z)]\,d\sigma$, and it is known that $\int_0^{2\pi} |u(re^{i\theta})|\,d\theta$ increases to the limit $\pi \int_{|\eta|=1} d|\sigma|$ as $r \to 1$. If $f \in V_k$, then $k\pi$ is a bound in a limiting sense for the variation of the tangent angle at the boundary.

The characterization of domains of bounded boundary rotation has been studied by V. Paatero [P1].

In order to show that functions of bounded boundary rotation are close-to-convex of some order, we shall use the following theorem.

THEOREM 2.25. If μ_1 and μ_2 are probability measures on $|\eta| = 1$, then there exists a constant c, $|c| = 1$, such that

$$\exp\{\int_{|\eta|=1} \log(1-\eta z)\,d(\mu_1 - \mu_2)\} \prec (1+cz)/(1-z) .$$

<u>Proof</u>. Assume first that the probability measure μ_j is generated by a continuous strictly increasing function g_j of $[0,2\pi]$ onto $[0,1]$, $j=1,2$. Then

$$I(z) = \int_{|\eta|=1} \log(1-\eta z)\,d(\mu_1-\mu_2) = \int_0^{2\pi} \log(1-e^{i\theta}z)\,d[g_1(\theta)-g_2(\theta)]$$
$$= \int_0^1 \log[(1-e^{ig_1^{-1}(x)}z)/(1-e^{ig_2^{-1}(x)}z)]\,dx .$$

Since $\log[(1-e^{i\alpha}z)/(1-e^{i\beta}z)]$ maps U onto a horizontal strip of width π,

$$\sup_{z\in U} \mathfrak{Im}\, I(z) - \inf_{z\in U} \mathfrak{Im}\, I(z) \le \int_0^1 [\sup_U \arg(\ ) - \inf_U \arg(\ )]dx \le \pi .$$

Therefore $I(U)$ is contained in a horizontal strip of width π containing $I(0)=0$. The types of measures considered are dense in the weak topology on the set of probability measures on $|\eta|=1$. Therefore by a normal family argument $I(U)$ is contained in a horizontal strip of width π containing $I(0)=0$ for general probability measures μ_1,μ_2. So there exists a constant c, $|c|=1$, such that $\int_{|\eta|=1}\log(1-\eta z)\,d(\mu_1-\mu_2) \prec \log[(1+cz)/(1-z)]$, and the result follows by exponentiation.

<u>THEOREM 2.26</u>([B8]). For $k>2$, $V_k \subset C(\frac{1}{2}k-1)$; that is, functions with boundary rotation at most k are close-to-convex of order $\frac{1}{2}k-1$.

<u>Proof</u>. Let $f\in V_k$. Since $V_k \subset V_{k'}$ whenever $k<k'$, we may assume that

$$f'(z) = \exp\{\int_{|\eta|=1}\log(1-\eta z)\,d\sigma\} \quad \text{where} \quad \int_{|\eta|=1} d\sigma = -2 \quad \text{and} \quad \int_{|\eta|=1} d|\sigma| = k .$$

Then $\mu_1 = (|\sigma|+\sigma)/(k-2)$ and $\mu_2 = (|\sigma|-\sigma)/(k+2)$ are both probability measures on $|\eta| = 1$ and

$$\varphi(z) = \int_0^z \exp[-2 \int_{|\eta|=1} \log(1-\eta z) d\mu_2] dz \in V_2 = K .$$ Since $\sigma = (\tfrac{1}{2}k-1)(\mu_1-\mu_2) - 2\mu_2$, we have

$$\frac{f'(z)}{\varphi'(z)} = \exp\{(\tfrac{1}{2}k-1)\int_{|\eta|=1} \log(1-\eta z) d(\mu_1-\mu_2)\} .$$

By Theorem 2.25, $(f'/\varphi')^{1/(\frac{1}{2}k-1)}$ is subordinate to some half-plane mapping, and the result follows from the definition of $C(\tfrac{1}{2}k-1)$.

COROLLARY 2.27. Functions of bounded boundary rotation $k \le 4$ are close-to-convex, hence univalent.

Proof. For $k \le 4$, $V_k \subset C(\tfrac{1}{2}k-1) \subset C(1) = C$.

We now determine the closed convex hull of V_k for $k \ge 4$ in terms of $\overline{\text{co}}C(\beta)$, which is known from Theorem 2.22.

THEOREM 2.28 ([B8]). For $k \ge 4$, $\overline{\text{co}}V_k = \overline{\text{co}}C(\tfrac{1}{2}k-1)$.

Proof. Let σ be the signed measure on $|\eta| = 1$ with mass $-\tfrac{1}{2}k-1$ at $\eta = \eta_0$ and mass $\tfrac{1}{2}k-1$ at $\eta = \xi_0$ $(\xi_0 \ne \eta_0)$. Then $\int_{|\eta|=1} d\sigma = -2$ and $\int_{|\eta|=1} d|\sigma| = k$. So

$$k(z;\xi_0,\eta_0,\tfrac{1}{2}k-1) = \int_0^z (1-\xi_0 z)^{\frac{1}{2}k-1}/(1-\eta_0 z)^{\frac{1}{2}k+1} dz = \int_0^z \exp[\int_{|\eta|=1} \log(1-\eta z) d\sigma] dz \in V_k .$$

For $k \ge 4$, we therefore have $E_{\overline{\text{co}}C(\frac{1}{2}k-1)} \subset V_k$ by Theorem 2.22. By the Krein-Milman theorem $\overline{\text{co}}C(\tfrac{1}{2}k-1) \subset \overline{\text{co}}V_k$. On the other hand, $\overline{\text{co}}V_k \subset \overline{\text{co}}C(\tfrac{1}{2}k-1)$ by Theorem 2.26.

PROBLEM. Determine $\overline{\text{co}}V_k$ and $E_{\overline{\text{co}}V_k}$ for $2 < k < 4$. In many respects this is the more interesting case. One would expect similar results to the case $k \ge 4$.

We now solve the coefficient problem in $C(\beta)$ and V_k. The method is due to Brannan, Clunie, and Kirwan [B8,B7].

THEOREM 2.29. The coefficients of functions in $C(\beta)$, $\beta>0$, are dominated by the corresponding coefficients of $k(z;-1,1,\beta)\in C(\beta)$.

Proof. If $f\in C(\beta)$, then $f'=e^{i\alpha}p^{\beta}\varphi'$ where $\varphi\in K$ and $\Re e\, p>0$. By Theorem 2.11

$$f'(z)=e^{i\alpha}p(z)^{\beta}\int_{|\eta|=1}1/(1-\eta z)^2 d\mu=\sum_{j=0}^{\infty}(\eta z)^{2j}\int_{|\eta|=1}e^{i\alpha}p(z)^{\beta}(1+\eta z)/(1-\eta z)d\mu .$$

Now $e^{i\alpha/(\beta+1)}p(z)^{\beta/(\beta+1)}[(1+\eta z)/(1-\eta z)]^{1/(\beta+1)}\prec(1+cz)/(1-z)$ for some choice of c, $|c|=1$, depending on α, β, and p. Since $\beta+1>1$, the coefficients of $e^{i\alpha}p(z)^{\beta}(1+\eta z)/(1-\eta z)$ are dominated by the coefficients of $[(1+z)/(1-z)]^{\beta+1}$ by Theorem 2.21. Since μ is a probability measure and $|\eta^{2j}|=1$, the same is true of $\eta^{2j}\int_{|\eta|=1}e^{i\alpha}p(z)^{\beta}(1+\eta z)/(1-\eta z)d\mu$. Consequently, the coefficients of f' are dominated by the coefficients of

$$\sum_{j=0}^{\infty}z^{2j}[(1+z)/(1-z)]^{\beta+1}=(1+z)^{\beta}/(1-z)^{\beta+2}=\frac{d}{dz}k(z;-1,1,\beta),$$

and the result follows.

COROLLARY 2.30. The coefficients of functions in V_k for $k\geq 2$ are dominated by the corresponding coefficients of $k(z;-1,1,\tfrac{1}{2}k-1)\in V_k$.

Proof. Since $V_2=K$, the result is contained in Theorem 2.12 for $k=2$. For $k>2$ the result follows from Theorems 2.26 and 2.29. That $k(z;-1,1,\tfrac{1}{2}k-1)\in V_k$ was observed in the proof of Theorem 2.28.

CHAPTER 3. The Pólya - Schoenberg conjecture

The Hadamard product or convolution of $f(z) = \sum_{n=0}^{\infty} a_n z^n \in H(U)$ and $g(z) = \sum_{n=0}^{\infty} b_n z^n \in H(U)$ is

$$(f*g)(z) = \sum_{n=0}^{\infty} a_n b_n z^n \in H(U).$$

This product is associative, commutative, and distributive over addition. As an example,

$$f * 1/(1-\eta z) = f(\eta z), \qquad |\eta| \le 1,$$

so that $1/(1-z)$ is an identity for the product. If $f(0) = 0$, then

$$f * z/(1-\eta z) = f(\eta z)/\eta, \qquad |\eta| \le 1 .$$

Consequently, if $f \in K$, then $f * z/(1-\eta z) \in K$ for $|\eta| = 1$. Symbolically, then $K * E_{\overline{\text{co}}K} \subset K$. Actually, $K * E_{\overline{\text{co}}K} = K$ since $z/(1-z)$ is an identity for $*$ on K. One might be led to the conjecture:

PÓLYA-SCHOENBERG CONJECTURE ([P6]). The Hadamard product of convex mappings is again a convex mapping, i.e., $f,g \in K \Rightarrow f * g \in K$.

The product is a convolution in the sense that

$$(f*g)(z) = \frac{1}{2\pi i} \int_{|\zeta|=\rho<1} f(\zeta) g(z\zeta^{-1}) \zeta^{-1} d\zeta \quad \text{for} \quad |z| < \rho .$$

It is evident that $*$ is continuous in the product topology of $H(U) \times H(U)$.

We observe first that the conjecture is true with K replaced by its closed convex hull.

LEMMA 3.1 (Schur [S14]). If $p(z) = 1 + \sum_{n=1}^{\infty} p_n z^n$ and $q(z) = 1 + \sum_{n=1}^{\infty} q_n z^n$ belong to P, then $1 + \frac{1}{2}\sum_{n=1}^{\infty} p_n q_n z^n$ also belongs to P.

Proof. Fix $z \in U$. Then

$$0 < \frac{1}{2\pi} \int_{|\zeta|=\rho>|z|} \Re e\, p(z\zeta^{-1}) \cdot \Re e\, q(\zeta) \frac{d\zeta}{i\zeta} = \tfrac{1}{2}\Re e \frac{1}{2\pi i} \int_{|\zeta|=\rho} p(z\zeta^{-1})[q(\zeta) + \overline{q(\zeta)}]\frac{d\zeta}{\zeta}$$

$$= \tfrac{1}{2}\Re e[1 + \sum_{n=1}^{\infty} p_n q_n z^n + 1] .$$

THEOREM 3.2. If $f, g \in \overline{co}K$, then $f * g \in \overline{co}K$.

Proof. If $f, g \in \overline{co}K$, then $2f/z - 1 = 1 + 2\sum_{n=1}^{\infty} a_{n+1} z^n$, $2g/z - 1 = 1 + 2\sum_{n=1}^{\infty} b_{n+1} z^n \in P$ by Theorem 2.11. By Lemma 3.1, $2(f*g)/z - 1 = 1 + 2\sum_{n=1}^{\infty} a_{n+1} b_{n+1} z^n \in P$, so that $f*g \in \overline{co}K$ by Theorem 2.11.

Alternate Proof. Since $*$ is continuous and distributes over addition,

$$\int z/(1-\xi z)\,d\mu * \int z/(1-\eta z)\,d\nu = \int [z/(1-\xi z) * z/(1-\eta z)]\,d(\mu \times \nu)$$

$$= \int z/(1-\xi\eta z)\,d(\mu \times \nu) \in \overline{co}K ,$$

using the integral representations of Theorem 2.11.

The first major result toward verifying the conjecture was due to T. J. Suffridge [S22]. He showed that the convolution of convex mappings is close-to-convex, hence univalent. The conjecture has recently been verified in full by St. Ruscheweyh and T. Sheil-Small [R4], and we shall present their development.

LEMMA 3.3. If $\varphi \in K$ and $|c_j| \le 1$, $j = 1,2,3$, then

$$\Re e \left\{ \frac{\varphi * z/[(1-c_1z)(1-c_2z)(1-c_3z)]}{\varphi * z/[(1-c_1z)(1-c_2z)]} \right\} > \frac{1}{2} \quad \text{in } U$$

and

$$\varphi * z/[(1-c_2z)(1-c_3z)] \ne 0 \quad \text{in } U-\{0\} .$$

Proof. Substitute c_jz, $j = 1,2,3$, into Theorem 2.8 for t,ζ,z, respectively. Then

$$\frac{1}{2} < \Re e \left[\frac{c_3}{c_3-c_2} \frac{c_2-c_1}{c_3-c_1} \frac{\varphi(c_3z)-\varphi(c_1z)}{\varphi(c_2z)-\varphi(c_1z)} - \frac{c_2}{c_3-c_2} \right]$$

$$= \Re e \left[\frac{c_3}{c_3-c_2} \frac{c_2-c_1}{c_3-c_1} \frac{\varphi * [1/(1-c_3z)-1/(1-c_1z)]}{\varphi * [1/(1-c_2z)-1/(1-c_1z)]} - \frac{c_2}{c_3-c_2} \right]$$

$$= \Re e \left[\frac{c_3}{c_3-c_2} \frac{\varphi * z/[(1-c_1z)(1-c_3z)]}{\varphi * z/[(1-c_1z)(1-c_2z)]} - \frac{c_2}{c_3-c_2} \right]$$

$$= \Re e \frac{\varphi * \{c_3z/[(1-c_1z)(1-c_3z)] - c_2z/[(1-c_1z)(1-c_2z)]\}}{(c_3-c_2)\varphi * z/(1-c_1z)(1-c_2z)]}$$

$$= \Re e \frac{\varphi * z/[(1-c_1z)(1-c_2z)(1-c_3z)]}{\varphi * z/[(1-c_1z)(1-c_2z)]} .$$

If $c_1 = 0$, then $\frac{1}{2} < \Re e \frac{\varphi * z/[(1-c_2z)(1-c_3z)]}{\varphi(c_2z)/c_2}$, so that

$\varphi * z/[(1-c_2z)(1-c_3z)] \ne 0$ for $z \ne 0$.

LEMMA 3.4. Suppose $f \in H(U)$, is continuous in $\bar{U}$, $f(0) = 0$, and

$$\Im m\, f(e^{i\theta}) \begin{matrix} \ge 0 & \text{for} & \theta \in (\alpha,\beta) \\ \le 0 & \text{for} & \theta \in (\beta,\alpha+2\pi) \end{matrix} \qquad 0 \le \beta-\alpha < 2\pi .$$

Then $\Im m\{e^{\frac{1}{2}i(\alpha+\beta)}(1-e^{-i\alpha}z)(1-e^{-i\beta}z)f(z)/z\} \ge 0$ for all $z \in U$.

Proof. Since $f(0) = 0$, we obtain from the Poisson formula the representation

$$f(z) = \frac{i}{2\pi}\int_0^{2\pi}[(e^{i\theta}+z)/(e^{i\theta}-z)]\,\mathscr{I}m\, f(e^{i\theta})d\theta$$

$$= \frac{i}{2\pi}\int_0^{2\pi}[(e^{i\theta}+z)/(e^{i\theta}-z)-(e^{i\alpha}+z)/(e^{i\alpha}-z)]\,\mathscr{I}m\, f(e^{i\theta})d\theta$$

$$= \frac{iz}{\pi}\int_\alpha^{\alpha+2\pi}[(e^{i\alpha}-e^{i\theta})\,\mathscr{I}m\, f(e^{i\theta})]/[(e^{i\theta}-z)(e^{i\alpha}-z)]d\theta \,.$$

Therefore $\mathscr{I}m\{e^{\frac{1}{2}i(\alpha+\beta)}(1-e^{-i\alpha}z)(1-e^{-i\beta}z)f(z)/z\}$

$$= \frac{1-|z|^2}{\pi}\int_\alpha^{\alpha+2\pi}[\sin\tfrac{1}{2}(\theta-\alpha)][\sin\tfrac{1}{2}(\beta-\theta)]\,\mathscr{I}m\, f(e^{i\theta})/|e^{i\theta}-z|^2 d\theta \geq 0 \,.$$

<u>REMARK</u>. It follows from Fatou's theorem and the boundary correspondence that functions in $H_u(U)$ have nontangential boundary values a.e. Therefore the hypothesis of Lemma 3.4 makes sense a.e. for $f \in H_u(U)$ if we drop the continuity on $\overline{U}$. In this case the conclusion remains true, or else f is of the form $cg(\eta z)$ for some $c > 0$, $|\eta| = 1$, and $g \in E_{T_{\mathbb{R}}}$ (Hengartner and Schober [H3]).

<u>LEMMA 3.5</u>. Let $g \in S^*$ and $\alpha \in \mathbb{R}$. Then there exist $\beta, \gamma \in \mathbb{R}$ such that

$$\mathscr{I}m\{e^{i\gamma}(1-e^{-i\alpha}z)(1-e^{-i\beta}z)g(z)/z\} > 0, \quad z \in U \,.$$

<u>Proof</u>. Fix $\alpha \in \mathbb{R}$ and $g \in S^*$. Then $\sigma = \lim_{r\to 1} \arg g(re^{i\alpha})$ exists by Corollary 2.14. Since $g(U)$ is starlike with respect to the origin, $s = g(U) \cap \{re^{i\sigma} : -\infty < r < \infty\}$ is a line segment (possibly infinite) whose endpoints are accessible boundary points of $g(U)$. Therefore $g^{-1}(s)$ is a Jordan arc in U with endpoints $e^{i\alpha}$ and, say, $e^{i\beta}$, $0 \leq \beta-\alpha < 2\pi$. For $r < 1$ each $g(|z| \leq r)$ is starlike (strictly, since if $\arg g(re^{i\theta})$ is constant for $\theta \in [\theta_1, \theta_2]$, then $g \equiv$ constant). Therefore $g^{-1}(s)$ intersects the circle $|z| = r$ in exactly two points $re^{i\alpha(r)}$, $re^{i\beta(r)}$. We

apply Lemma 3.4 to $e^{-i\sigma}g(rz)$. Then $\mathcal{I}m\{e^{\frac{1}{2}i[\alpha(r)+\beta(r)]}(1-e^{-i\alpha(r)}z)$ $(1-e^{-i\beta(r)}z)e^{-i\sigma}g(rz)/z\}$ is nonnegative. Let $r \to 1$. Then $\mathcal{I}m\{e^{i\gamma}(1-e^{-i\alpha}z)(1-e^{-i\beta}z)g(z)/z\} \geq 0$ for $\gamma = \frac{1}{2}(\alpha+\beta)-\sigma$. If strict inequality is not the case, then by the minimum principle the expression is a (nonzero) constant so that strict inequality can be obtained by perturbing the constant γ.

THEOREM 3.6. Let $\varphi \in K$, $g \in S^*$, and $F \in H(U)$ with $\mathcal{R}e\, F > 0$. Then

$$\mathcal{R}e\, \frac{\varphi * (Fg)}{\varphi * g} > 0 \quad \text{in } U .$$

Proof. Fix c and σ with $|c| = |\sigma| = 1$, and define $h(z) = [(1+c\sigma z)/(1-\sigma z)]g(z)$. We apply Lemma 3.5 with $e^{-i\alpha} = -c\sigma$. Then there exist $\beta, \gamma \in \mathbb{R}$ such that

$$0 < \mathcal{I}m\{e^{i\gamma}(1+c\sigma z)(1-e^{-i\beta}z)g(z)/z\} = \mathcal{I}m\{e^{i\gamma}(1-e^{-i\beta}z)(1-\sigma z)h(z)/z\} .$$

Therefore $(1-e^{-i\beta}z)(1-\sigma z)h(z)/z \prec (1+\epsilon z)/(1-z)$ for some ϵ, $|\epsilon| = 1$ By Theorem 2.20 (with $\alpha = 1$) there exists a probability measure μ such that

$$(1-e^{-i\beta}z)(1-\sigma z)h(z)/z = \int_{|\eta|=1} (1+\epsilon\eta z)/(1-\eta z)\,d\mu .$$

Now

$$\begin{aligned}\varphi * h &= \int_{|\eta|=1} \varphi * z(1+\epsilon\eta z)/[(1-e^{-i\beta}z)(1-\sigma z)(1-\eta z)]\,d\mu \\ &= \int_{|\eta|=1} \varphi * \{(1+\epsilon)z/[(1-e^{-i\beta}z)(1-\sigma z)(1-\eta z)] - \epsilon z/[(1-e^{-i\beta}z)(1-\sigma z)]\}\,d\mu \\ &= \{\varphi * z/[(1-e^{-i\beta}z)(1-\sigma z)]\}\left\{(1+\epsilon)\int_{|\eta|=1} \frac{\varphi * z/[(1-e^{-i\beta}z)(1-\sigma z)(1-\eta z)]}{\varphi * z/[(1-e^{-i\beta}z)(1-\sigma z)]}\,d\mu - \epsilon\right\}.\end{aligned}$$

The first factor is nonzero for $0 < |z| < 1$ by Lemma 3.3. The second factor can vanish only if

$$\mathrm{Re}\int_{|\eta|=1} \frac{\varphi * z/[(1-e^{-i\beta}z)(1-\sigma z)(1-\eta z)]}{\varphi * z/[(1-e^{-i\beta}z)(1-\sigma z)]}\, d\mu = \mathrm{Re}\ \epsilon/(1+\epsilon) = \tfrac{1}{2}\ ,$$

and this is impossible also by Lemma 3.3. Therefore

$$\varphi * [(1+c\sigma z)/(1-\sigma z)]g \neq 0 \quad \text{for} \quad 0<|z|<1 \quad \text{and} \quad |c| = |\sigma| = 1\ .$$

For $c=-1$ we have $\varphi * g \neq 0$ for $0<|z|<1$. By writing

$$\varphi * [(1+c\sigma z)/(1-\sigma z)]g = \tfrac{1}{2}(1+c)\varphi * [(1+\sigma z)/(1-\sigma z)]g + \tfrac{1}{2}(1-c)\varphi * g\ ,$$

we see that

$$G = \frac{\varphi * [(1+\sigma z)/(1-\sigma z)]g}{\varphi * g} \neq \frac{c-1}{c+1} \quad \text{for } |c|=1,\ c\neq -1, \text{ and } 0<|z|<1\ .$$

Now $G(0) = 1$, G is analytic in U, and the statement above implies that G assumes no value on the imaginary axis. By continuity then, $\mathrm{Re}\, G>0$ in U.

Finally, if $F \in H(U)$ and $\mathrm{Re}\, F>0$, then $F(z)/F(0) \prec (1+bz)/(1-z)$ for some constant b, $|b|=1$. By Theorem 2.20 (with $\alpha=1$) there exists a probability measure ν such that

$$F(z)/F(0) = \int_{|\sigma|=1} (1+b\sigma z)/(1-\sigma z)\, d\nu\ .$$

Now

$$\begin{aligned}(\varphi * Fg)/F(0) &= \int_{|\sigma|=1} \varphi * [(1+b\sigma z)/(1-\sigma z)]g\ d\nu \\ &= \tfrac{1}{2}(1+b)\int_{|\sigma|=1} \varphi * [(1+\sigma z)/(1-\sigma z)]g\ d\nu + \tfrac{1}{2}(1-b)\varphi * g\ ,\end{aligned}$$

and $(\varphi * Fg)/[F(0)\varphi * g] = \tfrac{1}{2}(1+b)H + \tfrac{1}{2}(1-b)$ is subordinate to $(1+bz)/(1-z)$ since $H = \int_{|\sigma|=1} G\, d\nu$ satisfies $\mathrm{Re}\, H>0$ and $H(0)=1$. Therefore $(\varphi * Fg)/[F(0)\varphi * g]$ takes U into the same half-plane that $F(z)/F(0)$ does. By a rotation, $\mathrm{Re}\{[\varphi * (Fg)]/(\varphi * g)\} > 0$.

The following theorem contains the affirmative solution of the Pólya-Schoenberg conjecture.

<u>THEOREM 3.7</u> (Ruscheweyh and Sheil-Small [R4]). Let $\varphi, \psi \in K$, $g \in S^*$, and $f \in C$. Then

$$\varphi * \psi \in K, \quad \varphi * g \in S^*, \quad \text{and} \quad \varphi * f \in C.$$

<u>Proof</u>. $F = zg'/g \in P$ by Theorem 2.3. Therefore

$$\text{Re}\, \frac{z(\varphi * g)'}{\varphi * g} = \text{Re}\, \frac{\varphi * (zg')}{\varphi * g} = \text{Re}\, \frac{\varphi * (Fg)}{\varphi * g} > 0$$

by Theorem 3.6. Consequently, $\varphi * g \in S^*$ by Theorem 2.7(a).

By Theorems 2.4(d) and 2.7, $z\psi' \in S^*$. Therefore $z(\varphi * \psi)' = \varphi * (z\psi') \in S^*$ by what has just been proved. Consequently, $\varphi * \psi \in K$ by Theorem 2.4(d).

By Corollary 2.5, $zf' = e^{i\alpha}\tilde{g} F$ where $\tilde{g} \in S^*$, $\alpha \in \mathbb{R}$, and $\text{Re}\, F > 0$. Therefore

$$\text{Re}\, \frac{z(\varphi * f)'}{e^{i\alpha}(\varphi * \tilde{g})} = \text{Re}\, \frac{\varphi * (zf')}{e^{i\alpha}(\varphi * \tilde{g})} = \text{Re}\, \frac{\varphi * (F\tilde{g})}{\varphi * \tilde{g}} > 0$$

by Theorem 3.6. Since $\varphi * \tilde{g} \in S^*$ by what has been proved earlier, $\varphi * f \in C$ by Corollary 2.5.

CHAPTER 4. Representation of continuous linear functionals

Let D be a domain, $D \subset \mathbb{C}$. Denote by $H'(D)$ the space of continuous linear functionals on $H(D)$. A function g is analytic on the closed set $\mathbb{C}-D$ if there exists an open set O_g containing $\mathbb{C}-D$ and an extension of g to O_g that is analytic in each component of O_g . We observe that if C is a finite system of rectifiable curves in $D \cap O_g$, then $L(f) = \frac{1}{2\pi i}\int_C f(z)g(z)dz$ defines a continuous linear functional on $H(D)$. The converse is also true:

THEOREM 4.1 (Caccioppoli [C1]; see also Köthe [K5]). Let $L \in H'(D)$. Then there exists a function g analytic in $\mathbb{C}-D$, vanishing at ∞, and a finite system C of rectifiable Jordan curves in $D \cap O_g$ such that

$$L(f) = \frac{1}{2\pi i}\int_C f(z)g(z)dz \quad \text{for all} \quad f \in H(D) .$$

Proof. Let $\{D_n\}$ be an exhaustion of D (i.e., $\bigcup_{n=1}^{\infty} D_n = D$) such that for every n, D_n is a domain, $\overline{D}_n$ is compact, $\overline{D}_n \subset D_{n+1}$, and ∂D_n consists of finitely many rectifiable Jordan curves. Define $\|f\|_n = \sup_{\overline{D}_n}|f|$.

Now let $L \in H'(D)$. Suppose there exists a sequence $\{f_n\}$ in $H(D)$ with $\|f_n\|_n = 1$ such that $|L(f_n)| \to \infty$ as $n \to \infty$. The sequence $\{f_n\}$ is locally uniformly bounded, hence normal, so there exists a subsequence $f_{n_k} \to f \in H(D)$. On the one hand, $|L(f_{n_k})| \to |L(f)| < \infty$ since L is continuous; on the other hand, $|L(f_{n_k})| \to \infty$ by assumption. Consequently, there exist constants M and m such that $|L(f)| \le M$ whenever $\|f\|_m = 1$ and $f \in H(D)$.

So if we consider $H(D)$ as a subspace of the Banach space B

of continuous functions on $\overline{D}_m$ that are analytic in D_m with $\|\cdot\|_m$ as norm, then we have shown that L is a bounded linear functional on the subspace $H(D)$. By the Hahn-Banach theorem L can be extended to a continuous linear functional $\overline{L}$ on B.

For fixed $\zeta \in \mathbb{C}-\overline{D}_m$, the function $1/(\zeta-z) \in B$. Let $g(\zeta) = \overline{L}(1/(\zeta-z))$. If $\zeta_o \in \mathbb{C}-\overline{D}_m$, then

$$[g(\zeta)-g(\zeta_o)]/(\zeta-\zeta_o) = -\overline{L}(1/[(\zeta-z)(\zeta_o-z)]) \to -\overline{L}(1/(\zeta_o-z)^2)$$

as $\zeta \to \zeta_o$ since $\overline{L}$ is continuous. Therefore g is analytic in each component of the open set $O_g = \mathbb{C}-\overline{D}_m \supset \mathbb{C}-D$. Since $\overline{D}_m$ is compact, g is analytic in a neighborhood of ∞ and clearly vanishes at ∞.

Now let $C = \partial D_{m+1}$ be oriented positively with respect to D_{m+1}. Then $C \subset D \cap O_g$, and if $f \in H(D)$, then

$$f(z) = \frac{1}{2\pi i}\int_C f(\zeta)/(\zeta-z)d\zeta \quad \text{for all} \quad z \in D_{m+1} \supset \overline{D}_m$$

by Cauchy's formula. Therefore

$$L(f) = \overline{L}(f) = \frac{1}{2\pi i}\int_C f(\zeta)\overline{L}(1/(\zeta-z))d\zeta = \frac{1}{2\pi i}\int_C f(\zeta)g(\zeta)d\zeta .$$

The interchange of $\overline{L}$ and $\int_C$ is permitted since the integral is a uniform limit of partial sums and $\overline{L}$ is continuous on B. (Note $\inf |\zeta-z| > 0$ for $z \in \overline{D}_m$ and $\zeta \in C = \partial D_{m+1}$.)

Additional representations for $H'(D)$ can easily be given in case D is any disk, annulus, or finitely connected domain. We give the most elementary of these:

<u>COROLLARY 4.2</u> (Toeplitz [T1]). $L \in H'(U)$ iff

$$L(f) = L\Big(\sum_{n=0}^{\infty} a_n z^n\Big) = \sum_{n=0}^{\infty} a_n b_n$$

where $\limsup_{n\to\infty} |b_n|^{1/n} < 1$.

<u>Proof</u>. If $L \in H'(U)$, then by Theorem 4.1 there exists a $g(z) = \sum_{n=0}^{\infty} b_n/z^{n+1}$ convergent in $|z| > r$ for some $r < 1$ ($\Rightarrow \limsup_{n\to\infty} |b_n|^{1/n} \le r < 1$) such that

$$L(f) = L\Big(\sum_{m=0}^{\infty} a_m z^m\Big) = \frac{1}{2\pi i}\int_{\substack{|z|=\rho \\ r<\rho<1}} \Big(\sum_{m=0}^{\infty} a_m z^m\Big)\Big(\sum_{n=0}^{\infty} b_n/z^{n+1}\Big)dz = \sum_{n=0}^{\infty} a_n b_n .$$

Conversely, assume $\{b_n\}$ is a sequence with $\limsup_{n\to\infty} |b_n|^{1/n} = \sigma < 1$. Then $g(z) = \sum_{n=0}^{\infty} b_n/z^{n+1}$ is analytic in $|z| > \sigma$, and for $f(z) = \sum_{n=0}^{\infty} a_n z^n \in H(U)$,

$$L(f) = \frac{1}{2\pi i}\int_{\substack{|z|=R \\ \sigma<R<1}} f(z)g(z)dz = \sum_{n=0}^{\infty} a_n b_n$$

defines a continuous linear functional on $H(U)$.

<u>EXERCISE</u>. Obtain representations for $H'(D)$ corresponding to Theorem 4.2 in case (i) D is any disk

(ii) D is an annulus

(iii) D is a finitely connected domain.

It will be convenient to have the following less precise representation for functionals in $H'(D)$.

<u>COROLLARY 4.3</u>. $L \in H'(D)$ iff there exists a finite complex Borel measure μ with compact support $K \subset D$ such that

$$L(f) = \int_K f \, d\mu \qquad \text{for all } f \in H(D) .$$

__Proof__. In Theorem 4.1 the differential $g(z)dz$ generates a measure μ on the compact set C. The "if" part is obvious.

We call μ a __representing measure__ for L. Since the support of μ is just a subset of D, the integral $\int_K f d\mu$ can be used to extend the definition of $L(f)$ to all functions f that are integrable with respect to μ. Then, for example,

$$L(1/(\zeta-z)) = \int_K 1/(\zeta-z)d\mu(\zeta) \text{ and } L(1/(f-w)) = \int_K 1/[f(\zeta)-w]d\mu(\zeta), \quad f \in H(D),$$

define analytic functions of $z \in \mathbb{C}-K$ and $w \in \mathbb{C}-f(K)$, vanishing at ∞. We shall use the same symbol for $L \in H'(D)$ and the above extensions.

The measure μ (and its support K) representing $L \in H'(D)$ is not uniquely determined. However, we have the following:

__COROLLARY 4.4__. Suppose $L \in H'(D)$ and E is a compact set in D containing the support of a representing measure for L. If $L(1/(\zeta-z)) = 0$ for all $z \in \mathbb{C}-E$, then $L \equiv 0$ on $H(D)$.

__Proof__. In the proof of Theorem 4.1 choose the exhaustion $\{D_n\}$ so that $E \subset D_1$. Then the curves C are in $D-E$ and $g(z) = L(1/(z-\zeta)) = 0$ on C.

For future use we note the following extension of Corollary 4.4:

__LEMMA 4.5__. Suppose $L \in H'(D)$ and E is a compact set in D containing the support of a representing measure for L. If $g \in H_u(D)$ and $L(1/(g-w))=0$ for all $w \in \mathbb{C}-g(E)$, then $L \equiv 0$ on $H(D)$.

Proof. If μ is a representing measure for L with support in the compact set $E \subset D$ and $g \in H_u(D)$, then $\tilde{\mu} = \mu \circ g^{-1}$ is a measure supported in $g(E)$ and $\tilde{L}(\tilde{f}) = \int_{g(E)} \tilde{f} d\tilde{\mu}$ defines a continuous linear functional on $H(g(D))$. If $\tilde{L}(1/(\omega - w)) = 0$ for all $w \in C - g(E)$, then $\tilde{L}(\tilde{f}) = 0$ for every $\tilde{f} \in H(g(D))$ by Corollary 4.4. The lemma follows by a change of variables.

CHAPTER 5. Faber polynomials

DEFINITIONS. Let $\Sigma_r = \{g \in H_u(|z| > r) : g(z) = z + \sum_{n=0}^{\infty} b_n z^{-n}\}$ and $g \in \Sigma_r$. Then the functions $F_m(t)$ generated by

$$\log \frac{z}{g(z)-t} = \sum_{m=1}^{\infty} \frac{1}{m} F_m(t) z^{-m}$$

for z belonging to a neighborhood of ∞ (depending on t) are the Faber polynomials of g. The coefficients γ_{mn} generated by

$$\log \frac{g(z)-g(\zeta)}{z-\zeta} = \sum_{m,n=1}^{\infty} \gamma_{mn} z^{-m} \zeta^{-n}, \quad |z|, |\zeta| > r,$$

are the Grunsky coefficients of g. Observe that $\gamma_{mn} = \gamma_{nm}$.

THEOREM 5.1. Let $\{F_m\}$ be the Faber polynomials of $g \in \Sigma_r$. Then (a) $F_m(t) = t^m + \sum_{n=0}^{m-1} \beta_n^{(m)} t^n$ where $[g^{-1}(t)]^m = t^m + \sum_{n=-\infty}^{m-1} \beta_n^{(m)} t^n$ near ∞ (i.e., F_m actually is a monic polynomial of degree m and consists of the principal part of the expansion of $[g^{-1}]^m$ near ∞), and $F_m(t) = t^m + \sum_{n=1}^{m-1} \frac{m}{n} b_m^{(n)} t^n + F_m(0)$ where $[g(z)]^{-n} = z^{-n} + \sum_{m=n+1}^{\infty} b_m^{(n)} z^{-m}$.

Also (b) $F_m(g(\zeta)) = \zeta^m - m \sum_{n=1}^{\infty} \gamma_{mn} \zeta^{-n}$ for $|z| > r$, where $\{\gamma_{mn}\}$ are the Grunsky coefficients of g (i.e., $F_m(g(\zeta)) = \zeta^m + o(1)$ near ∞).

Proof (Schiffer [S6]). (b): Fix ζ and choose z sufficiently large that

$$\log \frac{g(z)-g(\zeta)}{z-\zeta} = \log \frac{g(z)-g(\zeta)}{z} + \log \frac{1}{1-\zeta/z} = -\sum_{m=1}^{\infty} \frac{1}{m} F_m(g(\zeta)) z^{-m} + \sum_{m=1}^{\infty} \frac{1}{m}\left(\frac{\zeta}{z}\right)^m .$$

So $\frac{1}{m} F_m(g(\zeta)) = \frac{1}{m} \zeta^m - \sum_{n=1}^{\infty} \gamma_{mn} \zeta^{-n}$.

(a): Fix t and choose z sufficiently large that

$$\log \frac{z}{g(z)-t} = \log \frac{z}{g(z)} + \log \frac{1}{1-t/g(z)} = \log \frac{z}{g(z)} + \sum_{n=1}^{\infty} \frac{1}{n}\left[\frac{t}{g(z)}\right]^n$$

$$= \sum_{m=1}^{\infty} \frac{1}{m} F_m(0) z^{-m} + \sum_{m=1}^{\infty} \sum_{n=1}^{\infty} \frac{t^n}{n} b_m^{(n)} z^{-m} .$$

So $F_m(t) = F_m(0) + \sum_{n=1}^{m-1} \frac{m}{n} b_m^{(n)} t^n + t^m$, and the second representation follows. In particular, F_m is a polynomial. By part (b), already proved, $F_m(g(\zeta)) = \zeta^m + o(1)$ near ∞. Therefore the polynomial $F_m(t) = [g^{-1}(t)]^m + o(1)$ near ∞, and the first representation follows.

<u>NOTE</u>. We actually proved that F_m is the only polynomial such that $F_m(g(\zeta)) = \zeta^m + o(1)$ near ∞. Hence it is characterized by this condition.

<u>DEFINITIONS</u>. Let $S_r = \{f \in H_u(|z| < r): f(z) = z + \sum_{n=2}^{\infty} a_n z^n\}$ and $f \in S_r$. Then the functions $F_m(t)$ generated by

$$\log \frac{f(z)}{z[1-tf(z)]} = \sum_{m=1}^{\infty} \frac{1}{m} F_m(t) z^m$$

for z belonging to a neighborhood of the origin (depending on t) are the <u>Faber polynomials</u> of f. The coefficients c_{mn} generated by

$$\log \frac{f(z)-f(\zeta)}{z-\zeta} = \sum_{m,n=0}^{\infty} c_{mn} z^m \zeta^n , \qquad |z|, |\zeta| < r ,$$

are the <u>Grunsky coefficients</u> of f. Observe that $c_{mn} = c_{nm}$.

<u>THEOREM 5.2</u>. Let $\{F_m\}$ be the Faber polynomials of $f \in S_r$. Then (a) $F_m(t) = t^m + \sum_{n=1}^{m-1} \frac{m}{n} a_m^{(n)} t^n + mc_{mo}$ where $[f(z)]^n = z^n + \sum_{m=n+1}^{\infty} a_m^{(n)} z^m$ (i.e., F_m actually is a monic polynomial of degree m),

and (b) $F_m(1/f(\zeta)) = \zeta^{-m} - m\sum_{n=1}^{\infty} c_{mn}\zeta^n$ for $|\zeta| < r$

(i.e., $F_m(1/f(\zeta)) = \zeta^{-m} + o(1)$ near the origin).

Here c_{mn} are the Grunsky coefficients of f.

Proof. If $f \in S_r$, then $g(z) = 1/f(1/z) \in \sum_{1/r}$. From the generating functions one sees that f and g have the same Faber polynomials. Therefore (a) follows from Theorem 5.1(a) by substitution since $b_m^{(n)} = a_m^{(n)}$ and $\sum_{m=1}^{\infty} \frac{1}{m} F_m(0) z^m = \log(1/[zg(1/z)]) = \log(f(z)/z)$ $= \sum_{m=1}^{\infty} c_{mo} z^m$ implies $F_m(0) = mc_{mo}$. Since

$$\log \frac{g(1/z)-g(1/\zeta)}{1/z-1/\zeta} = \log \frac{f(z)-f(\zeta)}{z-\zeta} - \log \frac{f(z)}{z} - \log \frac{f(\zeta)}{\zeta},$$

the Grunsky coefficients $\gamma_{mn} = c_{mn}$ for $m, n \geq 1$, and part (b) follows from Theorem 5.1(b).

EXERCISE. $g(z) = z + a$ maps the exterior of $|z| = r$ onto the exterior of the circle with center a and radius r. Show that its Faber polynomials are $F_n(t) = (t-a)^n$.

EXERCISE. $f(z) = z/[1-\bar{a}z/(R^2-|a|^2)]$ maps $|z| < R-|a|^2/R$ onto the disk with center a and radius R. Show that its Faber polynomials are $F_n(t) = [t+\bar{a}/(R^2-|a|^2)]^n$.

Note that the Faber polynomials for both the exterior and the interior of circles centered at the origin are just the powers of t.

EXERCISE. $f(z) = z/[1+z/(2p)]$ maps $|z| < 2|p|$ onto the half-plane containing the origin whose nearest boundary point is p. Show that its Faber polynomials are $F_n(t) = [t-1/(2p)]^n$.

<u>EXERCISE</u>. $g(z) = z + 1/(4z)$ maps $|z| > \frac{1}{2}$ onto the complement of the real interval $[-1,1]$ and $|z| > r$ for $r > \frac{1}{2}$ onto the exterior of ellipses with foci ± 1. Show that its Faber polynomials are

$$F_n(t) = 2^{-n}[(t+\sqrt{t^2-1})^n + (t-\sqrt{t^2-1})^n] .$$

These polynomials are also the Chebyshev polynomials of $[-1,1]$, i.e, the monic polynomials of given degree that deviate least from zero on $[-1,1]$.

We shall see that Faber polynomials may be used to represent analytic functions in a complementary region:

<u>THEOREM 5.3</u>. Let $g \in \Sigma_r$ and $\{F_n\}$ be its Faber polynomials. Suppose φ is analytic in the interior of $\gamma_R = g(|z| = R)$ for some $R > r$.

(a) Then φ has the representation

$$\varphi(t) = c_0 + \sum_{n=1}^{\infty} c_n F_n(t)$$

in the interior of γ_R.

(b) The coefficients $c_n = \frac{1}{2\pi i} \int_{|z|=\rho} \varphi(g(z)) z^{-n-1} dz$ where $r < \rho < R$ and $\limsup_{n\to\infty} |c_n|^{1/n} \le 1/R$.

(c) The convergence is uniform on compact subsets of the interior of γ_R.

(d) Moreover, the representation is unique.

(e) Conversely, if $\{c_n\}$ is a sequence with $\limsup_{n\to\infty} |c_n|^{1/n} \le 1/R$, then $c_0 + \sum_{n=1}^{\infty} c_n F_n(t)$ represents an analytic function in the interior of γ_R.

Proof. Let $r<\rho<R$. Then for t inside $\gamma_\rho = g(|z|=\rho)$

$$\varphi(t) = \frac{1}{2\pi i}\int_{\gamma_\rho} \frac{\varphi(\zeta)}{\zeta-t}d\zeta = \frac{1}{2\pi i}\int_{|z|=\rho} \frac{\varphi(g(z))}{g(z)-t}g'(z)dz$$

$$= \frac{1}{2\pi i}\int_{|z|=\rho} \varphi(g(z))\Big[\frac{1}{z}+\sum_{n=1}^{\infty}F_n(t)z^{-n-1}\Big]dz = c_0 + \sum_{n=1}^{\infty}c_nF_n(t)$$

where $c_n = \frac{1}{2\pi i}\int_{|z|=\rho}\varphi(g(z))z^{-n-1}dz$. Since the series for the generating function of the Faber polynomials converges uniformly for $|z|=\rho$ and t in a compact subset of the interior of γ_ρ, the interchange of integration and summation is justified, and the final series converges uniformly on compact subsets of the interior of γ_ρ.

We note that $\limsup_{n\to\infty}|c_n|^{1/n} \le \limsup_{n\to\infty}(\max_{|z|=\rho}|\varphi(g(z))|)^{1/n}/\rho = 1/\rho$. Parts (a)-(c) now follow by letting $\rho\to R$.

(d) If $c_0 + \sum_{n=1}^{\infty}c_nF_n(t) \equiv d_0 + \sum_{n=1}^{\infty}d_nF_n(t)$, then for $r<\rho<R$

$$0 = \frac{1}{2\pi i}\int_{|z|=\rho}\Big[(c_0-d_0) + \sum_{n=1}^{\infty}(c_n-d_n)F_n(g(z))\Big]z^{-m-1}dz$$

$$= \frac{1}{2\pi i}\int_{|z|=\rho}\Big[(c_0-d_0) + \sum_{n=1}^{\infty}(c_n-d_n)(z^n+o(1))\Big]z^{-m-1}dz = c_m - d_m$$

for every $m \ge 0$.

(e) If $\limsup_{n\to\infty}|c_n|^{1/n} \le 1/R$, then $C(z) = \sum_{n=0}^{\infty}c_nz^n$ converges for $|z|<R$. For each ρ, $r<\rho<R$, the function

$$\frac{1}{2\pi i}\int_{|z|=\rho}C(z)g'(z)/[g(z)-t]dz$$

is analytic for t inside $\gamma_\rho = g(|z|=\rho)$. Since the series for $C(z)$ converges uniformly on $|z|=\rho$,

$$\frac{1}{2\pi i}\int_{|z|=\rho}C(z)\frac{g'(z)}{g(z)-t}dz = \sum_{n=0}^{\infty}c_n\frac{1}{2\pi i}\int_{|z|=\rho}z^n\Big[\frac{1}{z}+\sum_{m=1}^{\infty}F_m(t)z^{-m-1}\Big]dz$$

$$= c_0 + \sum_{n=1}^{\infty}c_nF_n(t).$$

The series therefore converges to an analytic function in the interior of γ_ρ . Let $\rho \to R$ for the conclusion.

COROLLARY 5.4. Let D be the interior of an analytic Jordan curve C . Determine r so that a function $g \in \Sigma_r$ maps $|z| > r$ onto the exterior of C . Then $\varphi \in H(D)$ iff

$$\varphi(t) = c_0 + \sum_{n=1}^{\infty} c_n F_n(t)$$

where $\limsup_{n\to\infty} |c_n|^{1/n} \le r$ and $\{F_n\}$ are the Faber polynomials of g .

Proof. Since C is analytic, $g \in \Sigma_{r'}$ for some $r' < r$ and Theorem 5.3 applies.

THEOREM 5.5. Let $f \in S_r$ and $\{F_n\}$ be its Faber polynomials. Suppose ψ is analytic in the exterior of $\gamma_R = f(|z| = R)$ for some $R < r$ and finite at ∞ .

(a) Then ψ has the representation

$$\psi(t) = c_0 + \sum_{n=1}^{\infty} c_n F_n(1/t)$$

in the exterior of γ_R .

(b) The coefficients $c_n = \frac{1}{2\pi i} \int_{|z|=\rho} \psi(f(z)) z^{n-1} dz$, where $R < \rho < r$, and satisfy $\limsup_{n\to\infty} |c_n|^{1/n} \le R$.

(c) The convergence is uniform on compact subsets of the exterior of γ_R .

(d) Moreover, the representation is unique.

(e) Conversely, if $\{c_n\}$ is a sequence with $\limsup_{n\to\infty} |c_n|^{1/n} \le R$, then $c_0 + \sum_{n=1}^{\infty} c_n F_n(1/t)$ represents an analytic function in the exterior of γ_R , that is finite at ∞ .

Proof. Let $R<\rho<r$. Then for t outside $\gamma_\rho = f(|z|=\rho)$

$$\psi(t) = -\frac{1}{2\pi i}\int_{\gamma_\rho}\frac{\psi(\zeta)}{\zeta-t}d\zeta + \psi(\infty) = -\frac{1}{2\pi i}\int_{\gamma_\rho}\frac{\psi(\zeta)}{\zeta-t}d\zeta + \frac{1}{2\pi i}\int_{\gamma_\rho}\frac{\psi(\zeta)}{\zeta}d\zeta$$

$$= \frac{1}{2\pi i}\int_{\gamma_\rho}\frac{t\psi(\zeta)}{\zeta(t-\zeta)}d\zeta = \frac{1}{2\pi i}\int_{|z|=\rho}\psi(f(z))\frac{f'(z)}{f(z)[1-(1/t)f(z)]}dz$$

$$= \frac{1}{2\pi i}\int_{|z|=\rho}\psi(f(z))\Big[\frac{1}{z}+\sum_{n=1}^{\infty}F_n(1/t)z^{n-1}\Big]dz = c_0 + \sum_{n=1}^{\infty}c_nF_n(1/t)$$

where $c_n = \frac{1}{2\pi i}\int_{|z|=\rho}\psi(f(z))z^{n-1}dz$. We have chosen the counterclockwise orientation of $|z|=\rho$ and the corresponding orientation of γ_ρ. The remainder of the proof parallels the proof of Theorem 5.3.

COROLLARY 5.6. Let C be an analytic Jordan curve with exterior $\tilde{D}$ and interior D. Assume $0 \in D$, and determine r so that a function $f \in S_r$ maps $|z|<r$ onto D. Then $\psi \in H(\tilde{D})$ and is finite at ∞ iff

$$\psi(t) = c_0 + \sum_{n=1}^{\infty}c_nF_n(1/t)$$

where $\limsup_{n\to\infty}|c_n|^{1/n} \le r$ and $\{F_n\}$ are the Faber polynomials of f.

Proof. Since C is analytic, $f \in S_{r'}$ for some $r'>r$ and Theorem 5.5 applies.

There has been work concerning the representation of analytic functions in the interior of a nonanalytic Jordan curve by series of Faber polynomials. For references see the expository article of J. Curtiss [C4].

If D is a simply connected domain (properly contained in $\mathbb{C}$), then D is conformally equivalent to the unit disk U. Extremal problems over subsets of $H(D)$ can therefore be referred back, in theory, to extremal problems over corresponding subsets of $H(U)$. However, one can take into account properties of D by representing the functionals in terms of D itself. One such possibility is the following:

<u>THEOREM 5.7.</u> Let D and $\tilde{D}$ be the interior and exterior, respectively, of an analytic Jordan curve C, and $0 \in D$. Let $\{F_n\}$ be the Faber polynomials of a g in some $\Sigma_{\tilde{r}}$ mapping onto $\tilde{D}$ and $\{\tilde{F}_n\}$ be the Faber polynomials of an f in some S_r mapping onto D . Then $L \in H'(D)$ iff

$$L(\varphi) = L\Big(a_0 + \sum_{m=1}^{\infty} a_m F_m(t)\Big) = \sum_{m=0}^{\infty} a_m \beta_m$$

where $\beta_m = \sum_{n=0}^{\infty} b_n \epsilon_{mn}$, $\epsilon_{mn} = \frac{1}{2\pi i}\int_{|t|=R} F_m(t)\tilde{F}_{n+1}(1/t)\,dt$, $\limsup_{m\to\infty} |a_m|^{1/m} \le 1/\tilde{r}$, and $\limsup_{n\to\infty} |b_n|^{1/n} < r$.

<u>Proof</u>. By Corollary 5.4, $\varphi \in H(D)$ iff $\varphi(t) = \sum_{m=0}^{\infty} a_m F_m(t)$ where $\limsup_{m\to\infty} |a_m|^{1/m} \le 1/\tilde{r}$. By Corollary 5.6, ψ is analytic on the closed set $C-D$ and vanishes at ∞ iff $\psi(t) = \sum_{n=0}^{\infty} b_n[\tilde{F}_{n+1}(\frac{1}{t}) - \tilde{F}_{n+1}(0)]$ where $\limsup_{n\to\infty} |b_n|^{1/n} < r$. Choosing $\gamma_\rho = f(|z|=\rho)$ sufficiently close to C that ψ is analytic outside and on γ_ρ ,

$$L(\varphi) = \frac{1}{2\pi i}\int_{\gamma_\rho} \varphi(t)\psi(t)\,dt = \sum_{m,n=0}^{\infty} a_m b_n \frac{1}{2\pi i}\int_{\gamma_\rho} F_m(t)[\tilde{F}_{n+1}(1/t) - \tilde{F}_{n+1}(0)]\,dt = \sum_{m=0}^{\infty} a_m \beta_m$$

defines a continuous linear functional on $H(D)$. By Theorem 4.1 every $L \in H'(D)$ has such a representation. Here $\beta_m = \sum_{n=0}^{\infty} b_n \epsilon_{mn}$ and

$$\epsilon_{mn} = \frac{1}{2\pi i}\int_{\gamma_\rho} F_m(t)[\tilde{F}_{n+1}(1/t) - \tilde{F}_{n+1}(0)]\,dt = \frac{1}{2\pi i}\int_{|t|=R} F_m(t)\tilde{F}_{n+1}(1/t)\,dt$$

since F_m and $\tilde{F}_n$ are polynomials.

In case $D = U$, both $F_m(t) = t^m$ and $\tilde{F}_n(t) = t^n$; therefore the matrix $[\varepsilon_{mn}]$ reduces to the identity, and Theorem 5.7 reduces to Corollary 4.2. In the general case we may interpret Theorem 5.7 to say that all continuous linear extremal problems over compact subsets of $H(D)$ are just problems on (finite or infinite) linear combinations of coefficients.

PROBLEM. Suppose D is bounded by an analytic Jordan curve. Solve linear extremal problems over compact subsets of $H_u(D)$. For example, if $F_1(t) + \sum_{n=2}^{\infty} a_n F_n(t) \in H_u(D)$, what can be said about the coefficients a_n?

CHAPTER 6. Extremal length and equicontinuity

At this point one of our goals is to show compactness of a certain normalized family of univalent functions. The standard method depends on an area inequality and is developed in Appendix B. However, in order to give an idea of some of the basic tools that are useful for studying quasiconformal mappings, with which we shall be concerned later, we shall rather employ length-area principles.

It will be convenient to use the extended plane (Riemann sphere) $\bar{C} = C \cup \{\infty\}$ with the topology induced by the spherical metric

$$\sigma(z,\zeta) = \begin{cases} |z-\zeta| / \sqrt{(1+|z|^2)(1+|\zeta|^2)} & z,\zeta \in C \\ 1/\sqrt{1+|z|^2} & z \in C,\ \zeta = \infty . \end{cases}$$

The relative topology in C of course agrees with the familiar euclidean topology.

DEFINITION. Let Γ be a family of arcs in $\bar{C}$ and ρ a nonnegative Borel measurable function on C. We say ρ is admissible for Γ, and write $\rho \wedge \Gamma$, if $1 \le \int_\gamma \rho ds \le \infty$ for every locally rectifiable $\gamma \in \Gamma$. The (possibly improper) integral is with respect to arc length, and we allow $\rho \equiv \infty$ to be admissible. The modulus of Γ is

$$M(\Gamma) = \inf_{\rho \wedge \Gamma} \int_C \rho^2 dm$$

where m is Lebesgue measure on C. The extremal length of Γ is $1/M(\Gamma)$.

Since the extremal length and modulus are reciprocals, it is sufficient to study one or the other. We shall concentrate on the modulus.

<u>EXAMPLE</u>. Let $R = \{z : a < |z| < b\}$ and Γ_R be the family of arcs in R joining boundary components. If $\rho \wedge \Gamma_R$, then using the radial arcs

$$1 \le \left(\int_a^b \rho(r,\theta)\,dr\right)^2 = \left(\int_a^b \rho\sqrt{r}\,\frac{dr}{\sqrt{r}}\right)^2 \le \left(\int_a^b \rho^2 r\,dr\right)\log(b/a) \text{ for } 0 \le \theta \le 2\pi$$

and $2\pi/\log(b/a) = \int_0^{2\pi} 1/\log(b/a)\,d\theta \le \int_0^{2\pi}\int_a^b \rho^2 r\,dr\,d\theta \le \int_C \rho^2\,dm$. For $\rho = 1/[r\log(b/a)]$ in R and $\rho = 0$ outside R, one has $\rho \wedge \Gamma_R$ and equality in the above estimate. Therefore $M(\Gamma_R) = 2\pi/\log(b/a)$.

<u>EXAMPLE</u>. Let $R = \{z : a < |z| < b\}$ and $\tilde{\Gamma}_R$ be the family of closed curves in R separating boundary components. If $\rho \wedge \tilde{\Gamma}_R$, then using the concentric circles

$$1 \le \left(\int_0^{2\pi} \rho r\,d\theta\right)^2 \le 2\pi \int_0^{2\pi} \rho^2 r^2\,d\theta \quad \text{for } a < r < b$$

and

$$\frac{1}{2\pi}\log(b/a) = \frac{1}{2\pi}\int_a^b \frac{dr}{r} \le \int_a^b\int_0^{2\pi} \rho^2 r\,d\theta\,dr \le \int_C \rho^2\,dm .$$

For $\rho = 1/(2\pi r)$ in R and $\rho = 0$ outside R, one has $\rho \wedge \tilde{\Gamma}_R$ and equality in the above estimate. Therefore $M(\tilde{\Gamma}_R) = \frac{1}{2\pi}\log(b/a)$.

<u>REMARK</u>. Observe that $M(\tilde{\Gamma}_R) = 1/M(\Gamma_R)$.

<u>EXERCISE</u>. Let $Q = \{x+iy:\ 0 < x < a \text{ and } 0 < y < b\}$ and Γ_a, Γ_b be the families of arcs in the rectangle Q joining the sides of length a and b, respectively. Show that $M(\Gamma_a) = a/b = 1/M(\Gamma_b)$.

An arc γ_1 <u>contains</u> an arc γ_2 if γ_2 is the restriction of γ_1 under some parametrization.

<u>THEOREM 6.1</u>. The modulus has the following properties:

(1) If $\Gamma_1 \subset \Gamma_2$, then $M(\Gamma_1) \le M(\Gamma_2)$.

(2) If each $\gamma_1 \in \Gamma_1$ contains a $\gamma_2 \in \Gamma_2$, then $M(\Gamma_1) \le M(\Gamma_2)$.

(3) If $\Gamma = \Gamma_1 \cup \Gamma_2$, then $M(\Gamma) \le M(\Gamma_1) + M(\Gamma_2)$.

(4) If Γ_1 and Γ_2 are contained in disjoint domains and each $\gamma_j \in \Gamma_j$ contains a $\gamma \in \Gamma$ (j=1,2), then $M(\Gamma) \ge M(\Gamma_1) + M(\Gamma_2)$.

(5) If Γ_1 and Γ_2 are contained in disjoint domains and each $\gamma \in \Gamma$ contains a $\gamma_1 \in \Gamma_1$ and a $\gamma_2 \in \Gamma_2$, then $M(\Gamma)^{-1} \ge M(\Gamma_1)^{-1} + M(\Gamma_2)^{-1}$.

Proof. (1)-(2) If $\rho \wedge \Gamma_2$, then $\rho \wedge \Gamma_1$ so that $\inf_{\rho \wedge \Gamma_1} \le \inf_{\rho \wedge \Gamma_2}$.

(3) If $\rho_1 \wedge \Gamma_1$ and $\rho_2 \wedge \Gamma_2$, then $(\rho_1^2 + \rho_2^2)^{\frac{1}{2}} \wedge \Gamma$ so that

$$M(\Gamma) \le \int_C (\rho_1^2 + \rho_2^2)\, dm = \int_C \rho_1^2 dm + \int_C \rho_2^2 dm .$$

Take the infimum over ρ_1 and ρ_2.

(4) Suppose Γ_j is contained in the domain D_j $(j = 1,2)$, $D_1 \cap D_2 = \emptyset$, and χ_j is the characteristic function of D_j. Let $\rho \wedge \Gamma$. Then $(\rho\chi_j) \wedge \Gamma_j$ and

$$\int_C \rho^2 dm = \int_C (\rho\chi_1)^2 dm + \int_C (\rho\chi_2)^2 dm \ge M(\Gamma_1) + M(\Gamma_2)$$

since $\chi_1\chi_2 \equiv 0$. Take the infimum over all $\rho \wedge \Gamma$.

(5) Suppose $\rho_1 \wedge \Gamma_1$ and $\rho_2 \wedge \Gamma_2$. Without loss of generality, $\rho_1\rho_2 \equiv 0$ since Γ_1 and Γ_2 are contained in disjoint domains. Note that $[\lambda\rho_1 + (1-\lambda)\rho_2] \wedge \Gamma$ for $0 \le \lambda \le 1$. Set $A_j = (\int_C \rho_j^2 dm)^{-1}$ and choose $\lambda = A_1/(A_1 + A_2)$, $1-\lambda = A_2/(A_1 + A_2)$. Then

$$M(\Gamma)^{-1} \ge \{\int_C [\lambda\rho_1 + (1-\lambda)\rho_2]^2 dm\}^{-1} = \{\lambda^2 \int_C \rho_1^2 dm + (1-\lambda)^2 \int_C \rho_2^2 dm\}^{-1}$$

$$= \{A_1/(A_1 + A_2)^2 + A_2/(A_1 + A_2)^2\}^{-1} = A_1 + A_2 .$$

Take the infimum over all ρ_1 and ρ_2.

<u>REMARKS</u>. (1) and (2) say that arc families with more or shorter arcs have larger modulus. Actually, (1) is a special case of (3), and (2) is a special case of both (4) and (5). Parts (3) and (4) imply that the modulus is additive for disjoint families. In fact, (3)-(5) can easily be extended to countably many families. Then (3) and (1) imply that the modulus can be used to define an outer measure on the space of all arcs.

<u>EXERCISE</u>. If Γ is an arc family in $\overline{\mathbb{R}^n} = \mathbb{R}^n \cup \{\infty\}$ $(n \geq 2)$, then $\rho \wedge \Gamma$ is defined as before and the <u>p-modulus</u> $M_p(\Gamma) = \inf_{\rho\wedge\Gamma} \int_{\mathbb{R}^n} \rho^p dm$. Verify that properties (1)-(4) of Theorem 6.1 carry over directly to the p-modulus in n-space. Verify that property (5) extends to $M_p(\Gamma)^{1/(1-p)} \geq M_p(\Gamma_1)^{1/(1-p)} + M_p(\Gamma_2)^{1/(1-p)}$ for $p > 1$. (Hint: In the proof of (5) change all exponents -1 to $1/(1-p)$ and all exponents 2 to p.)

<u>EXERCISE</u>. If R is the domain between two concentric spheres in $\mathbb{R}^n$ of radii a and b $(a < b)$ and Γ_R is the family of arcs in R joining the two boundaries, show that the p-modulus

$$M_p(\Gamma_R) = \omega_{n-1}[\sigma(b^\sigma - a^\sigma)]^{1-p} \quad \text{for } p > 1,\ p \neq n, \text{ and}$$

$$M_n(\Gamma_R) = \omega_{n-1}[\log(b/a)]^{1-n} \quad \text{where } \omega_{n-1} = 2\pi^{\frac{1}{2}n}/\Gamma(\tfrac{1}{2}n)$$

is the surface area of the unit n-1 sphere in $\mathbb{R}^n$ and $\sigma = (p-n)/(p-1)$. Consider also the degenerate cases $a = 0$ and $b = \infty$ using property (2) of the previous exercise.

If $\Lambda: \mathbb{C} \to \mathbb{C}$ is a linear transformation, i.e., $\Lambda(z) = az + b\bar{z}$, define $\|\Lambda\| = \sup_{|z|=1} |\Lambda(z)| = |a| + |b|$. In particular, if $df(z)$ is the <u>differential</u> of f at z, then $\|df\| = |f_z| + |f_{\bar z}|$. We denote by J_f the <u>Jacobian determinant</u> $J_f = |f_z|^2 - |f_{\bar z}|^2$. f is of <u>class C^1</u> if its first order partial derivatives exist and are continuous.

<u>THEOREM 6.2</u>. Let f be a homeomorphism of a domain $D \subset \bar{C}$ into $\bar{C}$. Assume f is of class C^1 and $\|df\|^2 \le K|J_f|$ in $(D \cap C) - f^{-1}(\{\infty\})$ for some constant K. Then $K^{-1}M(\Gamma) \le M(f(\Gamma)) \le K\,M(\Gamma)$ for all arc families Γ in D.

<u>Proof</u>. Let $\tilde\rho \wedge f(\Gamma)$ and define $\rho = (\tilde\rho \circ f)\|df\|$ in $D_o = (D \cap C) - f^{-1}(\{\infty\})$ and zero otherwise. Then $\rho \wedge \Gamma$ since $1 \le \int_{f(\gamma)} \tilde\rho ds \le \int_\gamma (\tilde\rho \circ f)\|df\|ds = \int_\gamma \rho ds$. So

$$M(\Gamma) \le \int_C \rho^2 dm = \int_{D_o} (\tilde\rho \circ f)^2 \|df\|^2 dm \le K \int_{D_o} (\tilde\rho \circ f)^2 |J_f| dm \le K \int_C \tilde\rho^2 dm .$$

Taking the infimum over all $\tilde\rho \wedge f(\Gamma)$, we have the first inequality $M(\Gamma) \le K\,M(f(\Gamma))$.

The second inequality follows by applying the first one to f^{-1}. Indeed, the identities

$$(f^{-1})_w = \overline{f_z}/J_f , \quad (f^{-1})_{\bar w} = -f_{\bar z}/J_f$$

at corresponding points $w = f(z)$ imply

$$\|df^{-1}\| = \|df\|/|J_f| \quad \text{and} \quad J_{f^{-1}} = 1/J_f .$$

Consequently, $\|df^{-1}\|^2 \le K|J_{f^{-1}}|$ follows from $\|df\|^2 \le K|J_f|$.

<u>COROLLARY 6.3</u>. If f is a conformal mapping of a domain $D \subset \bar{C}$, then $M(f(\Gamma)) = M(\Gamma)$ for all arc families Γ in D.

Proof. If f is conformal, then $\|df\|^2 = |f'|^2 = J_f$ so that Theorem 6.2 holds with $K=1$.

Corollary 6.3 says that the modulus of arc families is a conformal invariant. We now define the class of mappings for which the modulus is a quasi-invariant.

DEFINITION. An orientation preserving homeomorphism f of a domain $D \subset \bar{C}$ into $\bar{C}$ is K-quasiconformal if

$$K^{-1} M(\Gamma) \le M(f(\Gamma)) \le K\, M(\Gamma) \quad \text{for all arc families } \Gamma \text{ in } D .$$

Corollary 6.3 implies that conformal mappings are 1-quasiconformal. The converse is also true.

DEFINITION. A domain $R \subset \bar{C}$ is a ring if $\bar{C}-R$ has exactly two components. We denote by Γ_R the family of all arcs in R joining these two components and by $\tilde{\Gamma}_R$ the family of all closed curves in R separating them.

COROLLARY 6.4. If R is a ring domain in $\bar{C}$, then $M(\Gamma_R) = 1/M(\tilde{\Gamma}_R)$.

Proof. The remark on page 49 carries over to arbitrary ring domains by the conformal invariance of Corollary 6.3. (For the degenerate annuli ($a=0$ or $b=\infty$, or both) in the examples preceding that remark, observe that $M(\Gamma_R) = 0$ and $M(\tilde{\Gamma}_R) = \infty$ by properties (2) and (1) of Theorem 6.1.)

LEMMA 6.5. If the ring domain $R \subset \bar{C}$ separates the points a_1, b_1 from the points a_2, b_2 and the spherical distance

$\sigma(a_j, b_j) \geq d > 0$, then

$$M(\Gamma_R) \geq 4d^2/\pi .$$

Proof. Let $\rho_o(z) = 1/[2d(1+|z|^2)]$. Then $\int_{\tilde{\gamma}} 1/(1+|z|^2)ds$ is the arc length of $\tilde{\gamma}$ in the spherical metric. If $\tilde{\gamma} \in \tilde{\Gamma}_R$, then $\tilde{\gamma}$ cannot separate a_j and b_j for either $j = 1$ or $j = 2$. So $\int_{\tilde{\gamma}} 1/(1+|z|^2)ds \geq 2d$ for all $\tilde{\gamma} \in \tilde{\Gamma}_R$. Therefore $\rho_o \wedge \tilde{\Gamma}_R$. By Corollary 6.4

$$M(\Gamma_R) = 1/M(\tilde{\Gamma}_R) \geq 1/\int_C \rho_o^2 dm = 4d^2/\pi .$$

Suppose $\mathfrak{X}$ is a topological space and $(\mathfrak{M}, \sigma)$ a metric space. A family $\mathfrak{F}$ of functions $f: \mathfrak{X} \to \mathfrak{M}$ is equicontinuous at a point $p_o \in \mathfrak{X}$ if for each $\varepsilon > 0$ there is a neighborhood N_o of p_o such that $\sigma(f(p), f(p_o)) < \varepsilon$ whenever $p \in N_o$ and $f \in \mathfrak{F}$. The neighborhood N_o must be the same for all $f \in \mathfrak{F}$. The family $\mathfrak{F}$ is equicontinuous on $\mathfrak{X}$ if it is equicontinuous at each point of $\mathfrak{X}$.

THEOREM 6.6. Let $\mathfrak{F}$ be a family of K-quasiconformal mappings of a domain $D \subset \bar{C}$. If each $f \in \mathfrak{F}$ omits two values in $\bar{C}$ that may depend on f but have spherical distance at least $d > 0$, then $\mathfrak{F}$ is an equicontinuous family in the spherical metric.

Proof. Let $z_o \in D$ and ε, $0 < \varepsilon < d$, be fixed. Choose $r > 0$ so that the disk $\{z: \sigma(z, z_o) < r\} \subset D$. Choose $\delta < r$ so that the ring $R = \{z: \delta < \sigma(z, z_o) < r\}$ has $M(\Gamma_R) < 4\varepsilon^2/(\pi K)$. We designate the disk $\{z: \sigma(z, z_o) < \delta\}$ by N_o and prove that $\sigma(f(z_1), f(z_o)) < \varepsilon$ whenever $z_1 \in N_o$ and $f \in \mathfrak{F}$.

Let $f \in \mathfrak{F}$. By assumption f omits two values a_f, b_f with $\sigma(a_f, b_f) \geq d$. For $z_1 \in N_o$ the ring $f(R)$ separates the points a_f, b_f from the points $f(z_1), f(z_o)$. By Lemma 6.5

$$M(f(\Gamma_R)) = M(\Gamma_{f(R)}) \geq \frac{4\eta^2}{\pi} \quad \text{where} \quad \eta = \min(\sigma(a_f, b_f), \sigma(f(z_1), f(z_o))).$$

Since f is K-quasiconformal, $M(f(\Gamma_R)) \leq K M(\Gamma_R) < 4\varepsilon^2/\pi$. Therefore $\eta < \varepsilon$. Because $\varepsilon < d \leq \sigma(a_f, b_f)$, η is $\sigma(f(z_1), f(z_o))$.

We now apply this result to a normalized family of conformal mappings. We shall return to quasiconformal mappings later.

<u>COROLLARY 6.7</u>. Let $z_o \in D \subset \mathbb{C}$. Then $\mathcal{S}(D, z_o) = \{f \in H_u(D) : f(z_o) = 0, f'(z_o) = 1\}$ is compact in the topology of $H(D)$.

<u>Proof</u>. The restricted family $\{f|_{D-\{z_o\}} : f \in \mathcal{S}(D, z_o)\}$ omits 0 and ∞. By Theorem 6.6 it is an equicontinuous family in the spherical metric with values in the compact space $\bar{\mathbb{C}}$. Normality follows then from Ascoli's theorem. By the maximum principle, uniform convergence on a sufficiently small circle about z_o implies uniform convergence inside also. Hence $\mathcal{S}(D, z_o)$ itself is normal in the spherical metric. Since $f(z_o) = 0$ and $f'(z_o) = 1$ for $f \in \mathcal{S}(D, z_o)$, infinite and constant limits, respectively, of locally uniformly convergent sequences are not possible; they must be univalent and have the same normalization. Hence $\mathcal{S}(D, z_o)$ is compact in the topology of $H(D)$.

<u>REMARKS</u>. Many of the topics in this chapter are valid in $\mathbb{R}^n$, $n \geq 2$. If in Theorem 6.2 one assumes that $\|df\|^n \leq K|J_f|$ and $\|df^{-1}\|^n \leq K|J_{f^{-1}}|$, then the result applies to the n-modulus. It

follows that the n-modulus is a conformal invariant. However, it is an unfortunate fact that the only conformal mappings of domains in $\mathbb{R}^n$ for $n \geq 3$ are Möbius mappings (compositions of translations, homothetic and orthogonal linear transformations, and inversions in spheres). For the definition of K-quasiconformal mapping one uses the quasi-invariance of the n-modulus. Then Theorem 6.6 carries over directly to $\mathbb{R}^n$; however, the proof of Lemma 6.5 must be replaced by other considerations. An excellent reference is J. Väisälä [V1].

CHAPTER 7. Compact families $\mathfrak{F}(D,\ell_1,\ell_2,P,Q)$ of univalent functions normalized by two linear functionals

Let D be a domain in $\mathbb{C}$. We shall be concerned with families of univalent functions in D that are normalized by two continuous linear functionals.

DEFINITION. For fixed $\ell_1,\ell_2 \in H'(D)$ and $P,Q \in \mathbb{C}$ define

$$\mathfrak{F} = \mathfrak{F}(D,\ell_1,\ell_2,P,Q) = \{f \in H_u(D) : \ell_1(f) = P \text{ and } \ell_2(f) = Q\} .$$

EXAMPLE. For fixed $z_o \in D$ and $\ell_1(f) = f(z_o)$, $\ell_2(f) = f'(z_o)$, the family $\mathfrak{F}(D,\ell_1,\ell_2,0,1)$ is just

$$\mathfrak{S}(D,z_o) = \{f \in H_u(D) : f(z_o) = 0,\ f'(z_o) = 1\}$$

of Corollary 6.7. The further restriction $D = U$, $z_o = 0$, gives the familiar schlicht class $S = H_u(U) \cap N$.

EXAMPLE. For fixed $p,q \in D$ and $\ell_1(f) = f(p)$, $\ell_2(f) = f(q)$, with $p \neq q$ and $P \neq Q$, the family $\mathfrak{F}(D,\ell_1,\ell_2,P,Q)$ consists of univalent functions in D with assigned values at two given points. We denote it by

$$\mathfrak{T}(D,p,q,P,Q) = \{f \in H_u(D): \ f(p) = P,\ f(q) = Q\} .$$

EXAMPLE. Suppose that D contains a neighborhood of ∞. Choose R sufficiently large that $\{|z| \geq R\} \subset D$, and let

$$\ell_1(f) = \frac{1}{2\pi i}\int_{|z|=R} f(z)/z^2\, dz\ ,\quad \ell_2(f) = \frac{1}{2\pi i}\int_{|z|=R} f(z)/z\, dz\ .$$

Then $\mathfrak{F}(D,\ell_1,\ell_2,1,0)$ is the family

$$\Sigma'(D) = \{f \in H_u(D): \ f(z) = z + o(1) \quad \text{as} \quad z \to \infty\} .$$

We define also the associated families

$$\Sigma(D) = \{f \in H_u(D): f(z) = z + O(1) \quad \text{as} \quad z \to \infty\},$$

which is of the form $\mathfrak{F}(D, \ell_1, \ell_1, 1, 1)$, and

$$\Sigma^0(D) = \{f \in \Sigma(D): 0 \notin f(D)\},$$

which is not of the prescribed form. If $D = \mathbb{C}$, then $\Sigma^0(\mathbb{C}) = \phi$, $\Sigma'(\mathbb{C})$ contains only the identity mapping, and $\Sigma(\mathbb{C})$ its translates. We discard this case from all future considerations. For $D = \{|z| > 1\}$ we abbreviate

$$\Sigma' = \Sigma'(|z| > 1), \quad \Sigma = \Sigma(|z| > 1), \text{ and } \Sigma^0 = \Sigma^0(|z| > 1).$$

In this chapter we shall discuss compactness of the families $\mathfrak{F}(D, \ell_1, \ell_2, P, Q)$. The content is joint work with W. Hengartner [H5].

It will be convenient to associate with $\ell_1, \ell_2 \in H'(D)$ and $P, Q \in \mathbb{C}$ two new functionals

$$\ell_o = \frac{1}{\ell_1(Q) - \ell_2(P)} [\ell_2(1)\,\ell_1 - \ell_1(1)\ell_2]$$

$$\tilde{\ell}_o = \frac{1}{\ell_1(Q) - \ell_2(P)} [Q\ell_1 - P\ell_2]$$

in case $\ell_1(Q) \neq \ell_2(P)$. Observe that we do not distinguish between the constants $P, Q, 1$ and the corresponding constant functions. Note that

$$\ell_o(1) = 0 \qquad \tilde{\ell}_o(1) = 1$$

and for $f \in \mathfrak{F}(D, \ell_1, \ell_2, P, Q)$

$$\ell_o(f) = -1 \qquad \tilde{\ell}_o(f) = 0 \,.$$

If $g \in H_u(D)$ and $\ell_o(g) \neq 0$, it is elementary to verify that

$$T(g) = [-1/\ell_o(g)][g - \tilde{\ell}_o(g)] \in \mathfrak{F}(D, \ell_1, \ell_2, P, Q) \,.$$

THEOREM 7.1. Let $\ell_1, \ell_2 \in H'(D)$ and $P, Q \in \mathbb{C}$. If

(a) $\ell_1(Q) \neq \ell_2(P)$ and

(b) $\ell_2(1)\,\ell_1(g) \neq \ell_1(1)\,\ell_2(g)$ for all $g \in H_u(D)$,

then the family $\mathfrak{F}(D, \ell_1, \ell_2, P, Q)$ is nonempty and compact.

Proof. By Corollary 6.7 the family $\mathfrak{S}(D, z_o)$ (z_o fixed in D) is compact, and nonempty since $(z - z_o) \in \mathfrak{S}(D, z_o)$. If (a) and (b) hold, then the associated functionals $\ell_o, \tilde{\ell}_o$ are defined and $\ell_o(g) \neq 0$ for all $g \in H_u(D)$. The transform T of $H_u(D)$, defined above, is continuous; therefore $T(\mathfrak{S}(D, z_o))$ is a nonempty compact subset of $\mathfrak{F}$. Let $f \in \mathfrak{F}$ and set $g = [f - f(z_o)]/f'(z_o) \in \mathfrak{S}(D, z_o)$. An elementary calculation shows that $T(g) = f$. Therefore T maps $\mathfrak{S}(D, z_o)$ onto $\mathfrak{F}$, and $T(\mathfrak{S}(D, z_o)) = \mathfrak{F}$ is nonempty and compact.

REMARKS. If one knows a priori that $\mathfrak{F} \neq \emptyset$, then (b) implies compactness. Indeed, (a) follows from (b) by substituting an $f \in \mathfrak{F}$.

That (a) alone is not sufficient for compactness is evident from the family

$$G = \{f \in H_u(U) : f(0) = 0,\ f''(0) = 2\},$$

which satisfies (a), but is not compact since $nz + z^2 \in G$ for every $n \geq 2$.

We have already observed the compactness of $\mathfrak{S}(D, z_o)$ in Corollary 6.7 and used it in the above proof. The compactness of $\mathfrak{T}(D, p, q, P, Q)$ also follows directly from Theorem 6.6. However, it is instructive to apply Theorem 7.1 in these cases:

EXAMPLE. For $\mathfrak{S}(D,z_o)$, (a) is the statement that $1 \neq 0$ and (b) is the statement that $0 \neq g'(z_o)$ for all $g \in H_u(D)$. Both are obviously correct.

EXAMPLE. For $\mathfrak{T}(D,p,q,P,Q)$, (a) is the statement that $Q \neq P$ and (b) is the statement that $g(p) \neq g(q)$ for all $g \in H_u(D)$. Again both are obviously true. We record the consequence:

COROLLARY 7.2. The family $\mathfrak{T}(D,p,q,P,Q)$ is compact.

In general, $\Sigma(D)$ is not compact since $g \in \Sigma(D)$ iff $g + c \in \Sigma(D)$ for every $c \in C$. For $\Sigma'(D)$, (a) is the statement that $0 \neq 1$ and (b) is the statement that $\frac{1}{2\pi i}\int_{|z|=R} g(z)/z^2 \, dz \neq 0$ for all $g \in H_u(D)$. Unfortunately, (b) is violated by $1/(z-c)$ whenever $c \in C-D$. However, we shall show in Theorem 7.5 that both $\Sigma'(D)$ and $\Sigma^0(D)$ are compact. We therefore turn to the question of how good are the conditions of Theorem 7.1.

DEFINITION. A domain D has a <u>strongly</u> <u>dense</u> <u>boundary</u> if the only degenerate (one point) components of $\bar{C} - g(D)$ are limits of nondegenerate components for each $g \in H_u(D)$.

Examples of domains with strongly dense boundaries are all simply connected domains with at least two boundary points and all finitely connected domains with no degenerate boundary components (in $\bar{C}$). In these cases $\bar{C} - g(D)$ has no degenerate components for each $g \in H_u(D)$.

We now have the following converse to Theorem 7.1.

THEOREM 7.3. Let $\ell_1, \ell_2 \in H'(D)$ and $P, Q \in C$. If the family $\mathfrak{F}(D, \ell_1, \ell_2, P, Q)$ is nonempty and compact, then

(a) $\ell_1(Q) \neq \ell_2(P)$.

If, in addition, D has a strongly dense boundary, then

(b) $\ell_2(1)\, \ell_1(g) \neq \ell_1(1)\, \ell_2(g)$ for all $g \in H_u(D)$.

Proof. Assume $\mathfrak{F} = \mathfrak{F}(D, \ell_1, \ell_2, P, Q) \neq \phi$, and fix an $f \in \mathfrak{F}$. If (a) does not hold, then the linear system

$$\ell_1(Af + B) = A\,P + B\,\ell_1(1) = P$$

$$\ell_2(Af + B) = A\,Q + B\,\ell_2(1) = Q$$

has rank at most one and is consistent since $A = 1$, $B = 0$, is a solution. Therefore $\mathfrak{F}$ cannot be compact.

Suppose now $\mathfrak{F} \neq \phi$, D has a strongly dense boundary, and (a) holds, but not (b). Then there exists a $g \in H_u(D)$ such that $\ell_o(g) = 0$. Let K be the support of a representing measure for ℓ_o. If the analytic function $\ell_o(1/(g-w))$ vanishes at each point of $C-g(D)$, then it vanishes identically in some open neighborhood N of $\bar{C}-g(D)$, $N \subset \bar{C}-g(D)$, since $\bar{C}-g(D)$ has no isolated points. Now $D-g^{-1}(N)$ is a compact subset of D containing K. By Lemma 4.5, $\ell_o \equiv 0$. However, this cannot be the case since ℓ_o is -1 on $\mathfrak{F}$. So there exists a point $w_o \in C-g(D)$ such that $\ell_o(1/(g-w_o)) \neq 0$.

If w_o belongs to a degenerate component of $C-g(D)$, then by the continuity of $\ell_o(1/(g-w))$ and the strongly dense boundary property we may replace it by a point on a nondegenerate component of $C-g(D)$. In any case, there is a point w_o on a nondegenerate component of $C-g(D)$ such that $\ell_o(1/(g-w_o)) \neq 0$.

By composing with the Schiffer boundary variations of Theorem C.3, there exist $g_n \in H_u(D)$ of the form

$$g_n = g + a_n/(g-w_o) + o(a_n) \quad \text{where } a_n \to 0$$

and the term $o(a_n)/a_n$ converges to zero uniformly on compact subsets of D as $n \to \infty$. Since $\ell_o(g) = 0$, we have

$$\ell_o(g_n) = a_n \ell_o(1/(g-w_o)) + o(a_n) \neq 0$$

for all n sufficiently large. At the same time,

$$f_n = T(g_n) = \frac{-1}{a_n \ell_o(1/(g-w_o))}[g - \tilde{\ell}_o(g)] + o(1)$$

belongs to $\mathfrak{F}$, and $\{f_n\}$ has no convergent subsequence. Therefore $\mathfrak{F}$ is not compact.

If D has a strongly dense boundary, then conditions (a) and (b) characterize the nonempty compact sets $\mathfrak{F}(D,\ell_1,\ell_2,P,Q)$. In particular, we have

<u>COROLLARY 7.4</u>. Let D be a finitely connected domain with no degenerate boundary components in $\bar{C}$, $\ell_1, \ell_2 \in H'(D)$, and $P, Q \in C$. Then $\mathfrak{F}(D,\ell_1,\ell_2,P,Q)$ is compact iff

(a) $\ell_1(Q) \neq \ell_2(P)$ and

(b) $\ell_2(1)\, \ell_1(g) \neq \ell_1(1)\, \ell_2(g)$ for all $g \in H_u(D)$.

We observed earlier that the family $\Sigma'(D)$ violates condition (b). In defining $\Sigma'(D)$, we assumed that ∞ was an isolated point of $\bar{C}-D$. Consequently, D does not have a strongly dense boundary. Therefore, by showing that $\Sigma'(D)$ is compact, we note that the additional hypothesis for Theorem 7.3(b) is not superfluous.

THEOREM 7.5. $\Sigma^0(D)$ and $\Sigma'(D)$ are compact. If $z + \sum_{n=0}^{\infty} b_n z^{-n} \in \Sigma^0$, then $|b_o| \le 2$. If $g \in \Sigma'$, then $g(|z|>1) \supset \{|w|>2\}$.

Proof. Let $\Delta = \{1/w : w \in D\} \cup \{0\}$. Then $g \in \Sigma^0(D)$ iff $f(z) = 1/g(1/z) \in \mathcal{S}(\Delta,0)$. Since $\mathcal{S}(\Delta,0)$ is compact by Corollary 6.7, it follows that $\Sigma^0(D)$ is compact.

In $\mathcal{S}(\Delta,0)$ the continuous functional $\frac{1}{2}|f''(0)|$ is bounded, say by B. If $g \in \Sigma^0(D)$, $g(z) = z + \sum_{n=0}^{\infty} b_n z^{-n}$ in a neighborhood of ∞, and $f(z) = 1/g(1/z)$, then $|b_o| = \frac{1}{2}|f''(0)| \le B$. Suppose now $h \in \Sigma'(D)$ and $w_o \notin h(D)$. Then $g = h - w_o \in \Sigma^0(D)$ and $|w_o| \le B$. Therefore $\Sigma'(D)$ is a subset of $\{g+c : g \in \Sigma^0(D), \ |c| \le B\}$. The latter set is compact since $\Sigma^0(D)$ is; hence its closed subset $\Sigma'(D)$ is also compact.

By Theorem B.2 (or the example on pp. 105-106) we may choose $B = 2$ if $\Delta = U$. Therefore $|b_o| \le 2$ for Σ^0 and $|w_o| \le 2$ for Σ'.

EXERCISE. If $g \in \Sigma^0$ with $|b_o| = 2$, show that $g(z) = z + 2\eta + \eta^2/z$ for some $|\eta| = 1$. If $g \in \Sigma'$ with $\max\{|w| : w \notin g(|z|>1)\} = 2$, show that $g(z) = z + \xi/z$, $|\xi| = 1$.

It is convenient at this point to conclude with the following observation:

THEOREM 7.6. If A is a compact subset of $H(D)$, then $\overline{\text{co}}\, A$ is also compact.

Proof. If A is a compact family, it is locally uniformly bounded. The locally uniform bounds apply also to $\overline{\text{co}}\, A$. Hence the closed set $\overline{\text{co}}\, A$ is also locally uniformly bounded, normal, and compact.

<u>COROLLARY 7.7</u>. If $\mathfrak{F}(D,\ell_1,\ell_2,P,Q)$ is compact, then $\overline{co}\ \mathfrak{F}(D,\ell_1,\ell_2,P,Q)$ is also compact.

CHAPTER 8. Properties of extreme points for some compact families $\mathfrak{F}(D,\ell_1,\ell_2,P,Q)$

In this chapter we shall study the extreme points for nonempty compact families $\mathfrak{F}(D,\ell_1,\ell_2,P,Q)$. In particular, we shall show that extreme points of the families $\mathfrak{S}(D,z_o)$ and $\Sigma(D,p,q,P,Q)$ are slit mappings and that the slits possess monotonicity properties. For the special case of the class S this leads to the remarkable property that f/z is univalent if f is an extreme point. By contrast, we shall see that the extreme points of Σ are too numerous to be of value.

If $\mathfrak{F}(D,\ell_1,\ell_2,P,Q)$ is nonempty and compact, it follows from Theorem 7.3(a) that the functionals ℓ_o and $\tilde{\ell}_o$ of Chapter 7 are defined and belong to $H'(D)$.

LEMMA 8.1. Let $\mathfrak{F} = \mathfrak{F}(D,\ell_1,\ell_2,P,Q)$ be compact and $f \in \mathfrak{F}$. Suppose there exist distinct finite points a,b belonging to a component of $\tilde{C}-f(D)$ and satisfying

$$\ell_o(\sqrt{(f-a)(f-b)}) \in (-1,1) .$$

Then there exist $\lambda \in (0,1)$ and distinct $f_1,f_2 \in \mathfrak{F}$ such that $C-\overline{f_j(D)} \neq \emptyset$ and

$$f = \lambda f_1 + (1-\lambda) f_2 .$$

In particular, $f \notin E_{\mathfrak{F}}$.

Proof. If a,b are finite, distinct, and belong to a component of $\tilde{C}-f(D)$, then $\psi(w) = \sqrt{(w-a)(w-b)}$ has a single-valued analytic branch in $f(D)$. Moreover, $w \pm \psi(w)$ are both univalent in $f(D)$;

in fact, in both cases $w = (\zeta^2 - ab)/(2\zeta - a - b)$ is the inverse function. It follows that $w \pm \psi(w)$ map $f(D)$ onto disjoint domains.

Let $\varphi_1 = f + (\psi \circ f)$ and $\varphi_2 = f - (\psi \circ f)$. Then $\varphi_1 \varphi_2 \in H_u(D)$ and have disjoint images. In addition $A\varphi_1 + B\varphi_2$ is never constant if $|A| + |B| \neq 0$ since $A[w + \psi(w)] + B[w - \psi(w)] = C$ has at most finitely many solutions. Therefore

$$f_j = T(\varphi_j) = [-1/\ell_0(\varphi_j)][\varphi_j - \tilde{\ell}_0(\varphi_j)]$$

are distinct for $j=1,2$, belong to $\mathfrak{F}$, and the complement of $f_j(D)$ contains an open set. If $\ell_0(\psi \circ f) \in (-1,1)$, then $\lambda = \frac{1}{2}[1 - \ell_0(\psi \circ f)] \in (0,1)$ and $f = \lambda f_1 + (1-\lambda) f_2$. To see the latter, observe first that $\lambda = -\frac{1}{2}\ell_0(\varphi_1)$ and $1 - \lambda = -\frac{1}{2}\ell_0(\varphi_2)$.

The following lemma gives conditions under which some of the hypotheses of Lemma 8.1 are satisfied.

LEMMA 8.2. Let $\mathfrak{F} = \mathfrak{F}(D, \ell_1, \ell_2, P, Q)$ be compact and $f \in \mathfrak{F}$. Then there exists in each neighborhood of ∞ and for each point a, $|a|$ sufficiently large, a Jordan curve C_a containing a and separating 0 and ∞, such that

$$\ell_0\left(\sqrt{(f-a)(f-b)}\right) \in (-1,1)$$

for all $b \in C_a$, $b \neq a$.

Proof. By Corollary 4.3 the functional ℓ_0 has a representing measure with compact support $K \subset D$. Let $\rho = \max_K |f|$, restrict $|a| > \rho$, and set $b = aw^2$. Then

$$\Psi_a(w) = \ell_0\left(\sqrt{(f-a)(f-b)}\right) = aw\, \ell_0\left(\sqrt{[1-f/a][1-f/(aw^2)]}\right)$$

is a single-valued analytic function of w in $\sqrt{\rho/|a|} < |w| < \infty$, and one has

$$\Psi_a(w) = \tfrac{1}{2}(w + 1/w) + O(1/a) \quad \text{as} \quad a \to \infty .$$

Furthermore, as $a \to \infty$ the functions Ψ_a converge locally uniformly on $\mathbb{C}-\{0\}$ to

$$\Psi_\infty(w) = \tfrac{1}{2}(w + 1/w),$$

which defines a univalent mapping of $\mathbb{C}-\{0\}$ onto the Riemann surface R consisting of two planes identified across the two intervals $[-1,1]$, denoted by I. Therefore for fixed ϵ, $0 < \epsilon < 1$, Ψ_a is a univalent mapping of $A_\epsilon = \{\epsilon < |w| < 1/\epsilon\}$ onto $\Psi_a(A_\epsilon) \subset R$, containing I, for all a with $|a|$ sufficiently large. Therefore Ψ_a^{-1} converges to Ψ_∞^{-1} and maps I onto a Jordan curve tending to the circle $|w| = 1$ as $a \to \infty$. Since Ψ_a is an odd function, $b = a\,(\Psi_a^{-1})^2$ maps both intervals $[-1,1]$ onto a Jordan curve C_a with the desired properties as $a \to \infty$.

Very little is known in general about extreme points of the families $\mathfrak{F}(D,\ell_1,\ell_2,P,Q)$. However, we can say the following:

THEOREM 8.3 (Hengartner and Schober [H6]). Suppose f is an extreme point of a compact family $\mathfrak{F}(D,\ell_1,\ell_2,P,Q)$. If the component of $\bar{\mathbb{C}}-f(D)$ containing ∞ is nondegenerate, then its restriction to some neighborhood of ∞ is a single Jordan arc with one endpoint at ∞. In particular, at most one component of $\mathbb{C}-f(D)$ is unbounded.

Proof. Assume that the component γ of $\bar{\mathbb{C}}-f(D)$ containing ∞ is nondegenerate. Then by Lemmas 8.1 and 8.2, there is a neighborhood of ∞ such that each finite point of γ in that neighborhood is a <u>cut point</u>, i.e., each finite point $p \in \gamma$ in that neighborhood has the property that $\gamma - \{p\}$ is not connected. The same lemmas imply that the point at ∞ is not a cut point of γ, for otherwise, there would be at least two points of γ on all Jordan curves that wind around ∞ in some neighborhood of ∞. By using the characterization of Jordan arcs as continua with just two points that are not cut points (cf. Hocking and Young [H7, Theorem 2-27]), it follows that the restriction of γ to some neighborhood of ∞ is a Jordan arc with one endpoint at ∞.

The following Theorem is due to L. Brickman [B9] in case $D = U$.

<u>THEOREM 8.4</u> ([B9],[H4]). Let $f \in E_{\mathcal{S}(D,z_0)}$. Then each component of $\bar{C} - f(D)$ is either a point or a Jordan arc that meets each circle about the origin in at most one point. In particular, at most one component of $C - f(D)$ can be unbounded.

<u>Proof</u>. For f in the class $\mathcal{S}(D,z_0)$ the functional $\ell_0(\psi \circ f) = -(\psi \circ f)'(z_0) = -\psi'(0)$. If some component of $\bar{C} - f(D)$ contains distinct points $a = Re^{i\alpha}$ and $b = Re^{i\beta}$ on $|w| = R$ and $\psi(w) = \sqrt{(w-a)(w-b)}$, then

$$[\ell_0(\psi \circ f)]^2 = [\psi'(0)]^2 = \cos^2[\tfrac{1}{2}(\alpha - \beta)] \in [0,1)$$

and $f \notin E_{\mathcal{S}(D,z_0)}$ by Lemma 8.1. For $f \in E_{\mathcal{S}(D,z_0)}$ it follows that nondegenerate components of $\bar{C} - f(D)$ must meet each circle about the origin in at most one point; thus they are Jordan arcs. Two components of $C - f(D)$ cannot be unbounded, for otherwise they would belong to a single component of $\bar{C}-f(D)$, in which there would be distinct points on all sufficiently large circles.

EXERCISE. Verify the assertion in the above proof that a nondegenerate closed connected set that meets each circle about the origin in at most one point is a Jordan arc (cf. [B9, p. 373] and [H7, Theorem 2-27]).

Denote by $[P,Q]$ the closed segment from P to Q.

THEOREM 8.5 ([H4]). Let $f \in E_{\Sigma}(D,p,q,P,Q)$. Then each component of $(\bar{C}-[P,Q])-f(D)$ is either a point or a Jordan arc that meets each ellipse with foci P,Q in at most one point. In particular, at most one component of $C-f(D)$ is unbounded.

Proof. If $f \in \Sigma(D,p,q,P,Q)$, then $\ell_o(\psi \circ f) = -[\psi(P)-\psi(Q)]/(P-Q)$. If some component of $(\bar{C}-[P,Q])-f(D)$ contains distinct points a,b on an ellipse with foci P,Q and eccentricity ϵ, then $a = \frac{1}{2}(P+Q) + \frac{1}{2}(P-Q)\cos(\alpha-i\delta)$ and $b = \frac{1}{2}(P+Q) + \frac{1}{2}(P-Q)\cos(\beta-i\delta)$ where $\cos\delta = 1/\epsilon$. So if $\psi(w) = \sqrt{(w-a)(w-b)}$, then $\psi^2(P) = \frac{1}{4}(P-Q)^2[1-\cos(\alpha-i\delta)][1-\cos(\beta-i\delta)]$ and $\psi^2(Q) = \frac{1}{4}(P-Q)^2[1+\cos(\alpha-i\delta)][1+\cos(\beta-i\delta)]$. Now one easily verifies the trigonometric identity

$$\left[\frac{\psi^2(P)+\psi^2(Q)}{(P-Q)^2} - \cos^2\left(\frac{\alpha-\beta}{2}\right)\right]^2 = \frac{4\psi^2(P)\psi^2(Q)}{(P-Q)^2}$$

so that

$$\left[\frac{\psi(P)\pm\psi(Q)}{P-Q}\right]^2 = \cos^2\left(\frac{\alpha-\beta}{2}\right)$$

for the correct choice of sign. To determine the correct sign, we recall that the component under consideration does not meet the segment $[P,Q]$. We may consider the left side as a continuous function of P, for fixed α,β, and let $P \to Q$ along the segment.

The right side is constant while the left side remains bounded only for the minus sign. Therefore $[\ell_0(\psi \circ f)]^2 = \cos^2[\frac{1}{2}(\alpha-\theta)] \in [0,1)$, and $f \notin E_{\Sigma}$ by Lemma 8.1. The rest of the proof follows as in the previous theorem or by appealing to Theorem 8.3.

Theorem 8.5 is a pleasing extension of Theorem 8.4. As $P,Q \to 0$, the ellipses with foci P,Q become circles about the origin.

We now restrict our attention to the familiar schlicht class $S = \mathcal{S}(U,0)$. If $f \in E_S$, Theorem 8.4 says that $\mathbb{C}-f(U)$ consists of a single arc tending monotonically to ∞. It was obtained from Lemma 8.1, which in effect gave a restriction on pairs of boundary points. It is natural to seek further geometric information involving more than two points.

THEOREM 8.6 (L. Brickman (unpublished)). Let R be any rational function of the form

$$1/R(z) = \sum_{j=1}^{n} t_j/(z-\alpha_j)$$

with $n \geq 2$, distinct $\alpha_1,\ldots,\alpha_n \in \mathbb{C}$, and $t_1,\ldots,t_n > 0$. Let $A = \{z : R'(z) = 0\}$. If $f \in S$ and $R(A) \subset \mathbb{C}-f(U)$, then $f \notin E_{\overline{\cos}}$.

Proof. For any fixed $w \in \mathbb{C}$ we may write the equation $R(z) = w$ in the form

$$z^n - (\alpha + tw)z^{n-1} + \ldots = 0 \quad \text{where} \quad \alpha = \sum_{k=1}^{n} \alpha_k \quad \text{and} \quad t = \sum_{k=1}^{n} t_k .$$

Consequently, the equation $R(z) = w$ has n solutions, and a solution has multiplicity greater than one iff it belongs to A. So if $w \notin R(A)$, then all roots of $R(z) = w$ are simple; that is, there exist n distinct numbers $\varphi_1(w),\ldots,\varphi_n(w)$ such that

$$(1)\quad R(\varphi_1(w)) = w, \ldots, R(\varphi_n(w)) = w$$

and $$(2)\quad \varphi_1(w) + \ldots + \varphi_n(w) = \alpha + tw .$$

For fixed $w \notin R(A)$ the function R is univalent in a neighborhood of each point $\varphi_1(w), \ldots, \varphi_n(w)$. Therefore R has n local inverse functions $\varphi_1, \ldots, \varphi_n$, which are defined and analytic in a neighborhood of w, and satisfy (1) and (2). If now $f \in S$ and $R(A) \subset \mathbb{C} - f(U)$, then each φ_j can be continued analytically throughout the simply connected region $f(U)$. Hence there are n analytic functions $\varphi_1, \ldots, \varphi_n$ satisfying (1) and (2) in $f(U)$. By (1) they are also univalent in $f(U)$, and by (2)

$$w = \sum_{j=1}^{n} [\varphi_j(w) - \varphi_j(0)]/t .$$

Also by (2), $\sum_{j=1}^{n} \varphi_j'(0)/t = 1$; so f is a convex combination of $f_j = [(\varphi_j \circ f) - \varphi_j(0)]/\varphi_j'(0) \in S$, $j = 1, \ldots, n$, once we verify that all $\varphi_j'(0) > 0$ (note that $t > 0$ by hypothesis). For this purpose, observe that $\varphi_j(0)$ is a zero of $R(z)$. Therefore $\varphi_j(0) = \alpha_k$ for some k, and

$$\varphi_j'(0) = 1/R'(\alpha_k) = \lim_{z \to \alpha_k} (z - \alpha_k)/R(z) = t_k > 0 .$$

Finally, all $f_j \neq f$; for otherwise, $[\varphi_j(w) - \varphi_j(0)]/\varphi_j'(0) \equiv w$ in $f(U)$ and $w \equiv R(\varphi_j(w)) = R(\varphi_j'(0)w + \varphi_j(0))$. However, the equation $w \equiv R(aw + b)$ implies $R(z) = (z-b)/a$. Consequently, f is a convex combination of n other functions of S. We conclude that $f \notin E_{\overline{co}S}$.

<u>REMARK</u>. By Corollary 7.7 the closed convex hull of S is compact; hence, $E_{\overline{co}S} \subset S$ by the Krein-Milman theorem (Appendix A). Therefore Theorem 8.6 gives implicit geometric restrictions on the

mapping properties of extreme points for $\overline{\text{co}}S$. If one is interested in linear extremal problems, then Theorem A.3 says that a complete knowledge of $E_{\overline{\text{co}}S}$ would be sufficient.

EXERCISE. In Theorem 8.6 with $n=2$, verify that $R(A)$ consists of two numbers of equal modulus and that any two numbers of equal modulus are possible. Conclude that $C-f(U)$ is a monotone arc if $f \in E_{\overline{\text{co}}S}$. (Actually, for $n=2$, Theorem 8.6 applies to $f \in E_S$.)

PROBLEM (L. Brickman). What additional information does Theorem 8.6 give about the monotone arc $C-f(U)$ for $f \in E_{\overline{\text{co}}S}$ when $n \geq 3$?

PROBLEM. What is the analogue of Theorem 8.6 for a general compact family $\mathfrak{F}(D,\ell_1,\ell_2,P,Q)$?

The previous theorems give geometric information about mapping properties of extreme points. It is of interest to convert the geometric properties into analytic ones. The following theorem is in this direction. If f is an extreme point of S, it concludes that both $\log(f/z)$ and f/z are univalent and have further geometric properties.

THEOREM 8.7 (Hengartner and Schober [H4]). Suppose $f \in S$ and $C-f(U)$ is a monotone arc (intersects each circle about the origin at most once).

Then (i) $\log(f/z) \in H_u(U)$ and maps U onto a domain D' that is convex in the v-direction (i.e., the intersection of D' with each vertical line is connected).

(ii) The vertical line with abcissa u meets $\overline{D'}$ in a segment of length $\ell_u < 2\pi$.

(iii) ℓ_u is a continuous strictly increasing function of u on (u_o, ∞) where $u_o = \inf\{\Re e\, w : w \in D'\}$, and $\lim_{u \to u_o} \ell_u = 0$, $\lim_{u \to \infty} \ell_u = 2\pi$.

(iv) $f/z \in H_u(U)$ and maps U onto a domain D'' with the property that each circle $|w| = r$ meets D'' in a single arc of ω_r radians. Moreover, ω_r is a continuous strictly increasing function of r on (r_o, ∞) where $r_o = \inf\{|w| : w \in D''\}$, and $\lim_{r \to r_o} \omega_r = 0$, $\lim_{r \to \infty} \omega_r = 2\pi$.

In particular, (i)-(iv) hold for $f \in E_S$.

<u>Proof</u>. (i): Suppose $f \in S$ and $C - f(U)$ is a monotone arc. Let f_n map U conformally onto $D_n = f(U) \cap \{|w| < n\}$ with $f_n(0) = 0$, $f'_n(0) > 0$. Then by a convergence theorem of T. Radó (see [G8, p. 59]), f_n converges to f uniformly on U relative to the spherical metric. We may extend f_n and f continuously to $\bar{U}$ in the spherical metric. Suppose then that $f(e^{i\theta_2}) = f_n(e^{i\theta_{2,n}})$ is the point of $C - f(U)$ nearest the origin, $f(e^{i\theta_1}) = \infty$, and $|f_n(e^{i\theta_{1,n}})| = n$ where $0 < \theta_2 - \theta_1 < 2\pi$, $0 < \theta_{2,n} - \theta_{1,n} < 2\pi$. Of course, the choice of $e^{i\theta_{1,n}}$ is not unique. Nevertheless, $e^{i\theta_{1,n}} \to e^{i\theta_1}$ and $e^{i\theta_{2,n}} \to e^{i\theta_2}$ as $n \to \infty$ by the uniform convergence.

For $\log(f_n/z)$ we have the Poisson representation

$$\log(f_n(z)/z) = \frac{1}{2\pi} \int_0^{2\pi} [\log|f_n(e^{i\theta})|][e^{i\theta} + z)/(e^{i\theta} - z)] d\theta .$$

So $[\log(f_n(z)/z]' = (1/\pi)\int_0^{2\pi}[\log|f_n(e^{i\theta})|][e^{i\theta}/(e^{i\theta}-z)^2]d\theta$

$$= \frac{1}{\pi i(e^{i\theta_{2,n}}-z)}\int_0^{2\pi}\log|f_n(e^{i\theta})|\ d\,\frac{e^{i\theta}-e^{i\theta_{2,n}}}{e^{i\theta}-z}$$

$$= \frac{i}{\pi(e^{i\theta_{2,n}}-z)}\int_0^{2\pi}\frac{e^{i\theta}-e^{i\theta_{2,n}}}{e^{i\theta}-z}\,d\log|f_n(e^{i\theta})|.$$

Since $\log|f_n(e^{i\theta})|$ is nonincreasing for $\theta_{1,n} < \theta < \theta_{2,n}$ and nondecreasing for $\theta_{2,n} < \theta < \theta_{1,n} + 2\pi$, it follows that

$$\Re\{-ie^{-\frac{1}{2}i(\theta_{1,n}+\theta_{2,n})}(e^{i\theta_{1,n}}-z)(e^{i\theta_{2,n}}-z)[\log(f_n(z)/z)]'\}$$

$$= \frac{-2(1-|z|^2)}{\pi}\int_{\theta_1}^{\theta_1+2\pi}\frac{\sin[\frac{1}{2}(\theta-\theta_{1,n})]\sin[\frac{1}{2}(\theta_{2,n}-\theta)]}{|e^{i\theta}-z|^2}\,d\log|f_n(e^{i\theta})|$$

is nonnegative. By passing to the limit, we have

$$\Re\{[\log(f/z)]'/\varphi'\} = \Re\{-ie^{-\frac{1}{2}i(\theta_1+\theta_2)}(e^{i\theta_1}-z)(e^{i\theta_2}-z)[\log(f/z)]'\} \geq 0$$

where $\varphi(z) = \frac{1}{2}\csc[\frac{1}{2}(\theta_2-\theta_1)]\log[(1-ze^{-i\theta_2})/(1-ze^{-i\theta_1})]$ is a mapping of U onto a horizontal strip. If strict inequality is not the case, then by the minimum principle $[\log(f/z)]' = i\beta\varphi'$ for some $\beta \in \mathbb{R}$, and $\log(f/z)$ is a univalent mapping of U onto a vertical strip, so that (i) is satisfied (we shall see in part (ii) that this case cannot occur). If strict inequality is the case, then $\log(f/z)$ is close-to-convex relative to φ. Therefore $g = \log(f/z)$ is univalent by Theorem 2.2. Since $u = \Re g(e^{i\theta}) = \log|f(e^{i\theta})|$ is continuous and decreasing for $\theta \in (\theta_1, \theta_2]$ and continuous and increasing for $\theta \in [\theta_2, \theta_1 + 2\pi)$, it follows that the image domain $D' = g(U)$ is convex in the v-direction.

(ii) and (iii): Since $\bar{\mathbb{C}}-f(U)$ is a Jordan arc, we may define a continuous decreasing function α of $[\theta_1, \theta_2]$ onto $[\theta_2, \theta_1 + 2\pi]$ such that $f(e^{i\alpha(\theta)}) = f(e^{i\theta})$. Then for $g = \log(f/z)$ we have

$$g(e^{i\theta}) - g(e^{i\alpha(\theta)}) = i[\alpha(\theta) - \theta] \quad \text{modulo} \quad 2\pi i \ .$$

Note that both sides are continuous functions of θ and vanish for $\theta = \theta_2$. We may therefore omit the "modulo $2\pi i$". Since the right

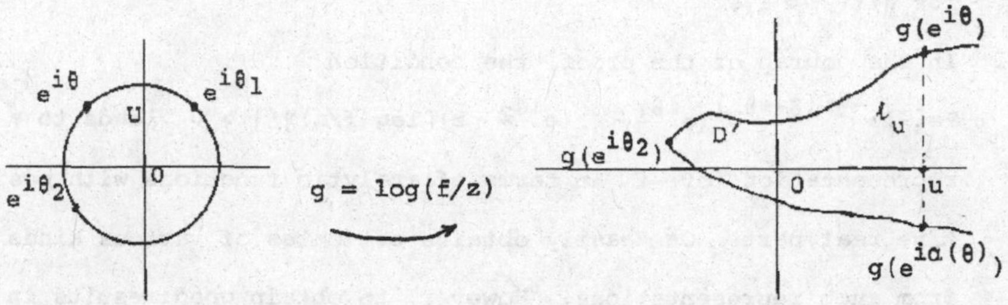

side is purely imaginary, $g(e^{i\theta})$ and $g(e^{i\alpha(\theta)})$ are endpoints of a vertical segment of length $\ell_u = \alpha(\theta) - \theta$, $u = \operatorname{Re} g(e^{i\theta})$. Moreover, $\alpha(\theta) - \theta$ is continuous and strictly decreasing on $[\theta_1, \theta_2]$; it is 2π at θ_1 and zero at θ_2. Since $u = \operatorname{Re} g(e^{i\theta}) = \log|f(e^{i\theta})|$ decreases as θ increases on $[\theta_1, \theta_2]$, parts (ii) and (iii) follow.

(iv): Since $\ell_u < 2\pi$ for all u, the exponential function is univalent in $\overline{D'}$. Consequently, $e^g = f/z \in H_u(U)$. The properties described in (iv) follow immediately from (ii) and (iii).

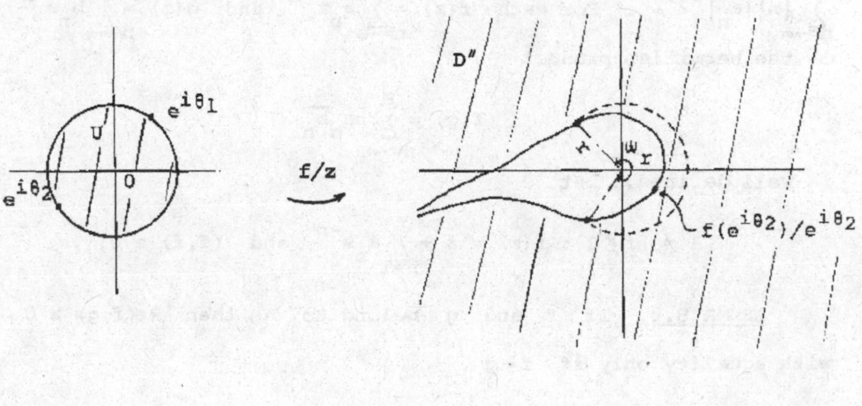

REMARKS. Suppose f satisfies the hypotheses of Theorem 8.7 (e.g., $f \in E_S$).

1. Since $f/z \neq 0$, univalent branches of $(f/z)^{1/p}$ can be defined for all $p \geq 1$.
2. In the course of the proof, the condition $\Re\{-ie^{-\frac{1}{2}i(\theta_1+\theta_2)}(e^{i\theta_1}-z)(e^{i\theta_2}-z)[\log(f/z)]'\} > 0$ leads to a representation for f in terms of analytic functions with positive real part. One easily obtains estimates of various kinds from such representations. However, to obtain good results in this direction one would also have to take into account properties (ii) and (iii).
3. Property (iv) implies that $[2/f''(0)][(f/z)-1] \in S$.

We have shown that the extreme points of S have valuable properties. By contrast, we close this chapter with a result of G. Springer [S17] (see also [K1]) to the effect that the extreme points of Σ' are too numerous to be of value.

Let $\mathfrak{a}$ denote the linear subspace of $H(|z|>1)$ consisting of functions $f(z) = \sum_{n=-\infty}^{\infty} a_n z^{-n}$ that satisfy the condition $\sum_{n=-\infty}^{\infty} |n||a_n|^2 < \infty$. For each $f(z) = \sum_{n=-\infty}^{\infty} a_n z^{-n}$ and $g(z) = \sum_{n=-\infty}^{\infty} b_n z^{-n}$ in $\mathfrak{a}$ the hermitian product

$$\langle f,g\rangle = \sum_{n=-\infty}^{\infty} n a_n \overline{b_n}$$

is well defined. Let

$$A = \{f \in \mathfrak{a} : f(z) = z + \sum_{n=1}^{\infty} a_n z^{-n} \text{ and } \langle f,f\rangle \geq 0\}.$$

<u>LEMMA 8.8</u>. If f and g belong to A, then $\Re\langle f,g\rangle \geq 0$, with equality only if $f = g$.

<u>Proof</u>. By the Cauchy-Schwarz inequality

$$\Re e\langle f,g\rangle = 1 - \Re e\sum_{n=1}^{\infty} n a_n \overline{b_n}$$
$$\geq 1 - \Big[\sum_{n=1}^{\infty} n|a_n|^2\Big]^{\frac{1}{2}}\Big[\sum_{n=1}^{\infty} n|b_n|^2\Big]^{\frac{1}{2}} \geq 0 .$$

If $\Re e\langle f,g\rangle = 0$, then $\sum_{n=1}^{\infty} n|a_n|^2 = 1 = \sum_{n=1}^{\infty} n|b_n|^2$, and by the condition for equality in the Cauchy-Schwarz inequality, $a_n = cb_n \quad (n \geq 1)$ for some $c > 0$. However, only $c = 1$ is compatible. Thus $f = g$.

<u>THEOREM 8.9</u> ([K1]). A is a closed convex subset of $H(|z|>1)$. Moreover,

$$E_A = \{f \in A : \langle f,f\rangle = 0\} .$$

<u>Proof</u>. Let $f_k(z) = z + \sum_{n=1}^{\infty} a_{n,k} z^{-n}$ belong to A, and suppose $f_k \to f$ locally uniformly in $|z| > 1$ as $k \to \infty$. Then $f(z) = z + \sum_{n=1}^{\infty} a_n z^{-n}$ where for each n, $\lim_{k\to\infty} a_{n,k} = a_n$. Since $\sum_{n=1}^{N} n|a_{n,k}|^2 \leq \sum_{n=1}^{\infty} n|a_{n,k}|^2 \leq 1$, by letting $k \to \infty$ we have $\sum_{n=1}^{N} n|a_n|^2 \leq 1$ for each N . Therefore $f \in A$, so that A is closed.

If $f = tf_1 + (1-t)f_2$ where $0 \leq t \leq 1$ and $f_1, f_2 \in A$, then

$$\langle f,f\rangle = t^2\langle f_1,f_1\rangle + 2t(1-t)\Re e\langle f_1,f_2\rangle + (1-t)^2\langle f_2,f_2\rangle \geq 0$$

by Lemma 8.8. Therefore $f \in A$ and A is convex. If $\langle f,f\rangle = 0$, then the three nonnegative terms must all be zero; in particular, $\Re e\langle f_1,f_2\rangle = 0$. In this case $f_1 = f_2$ by Lemma 8.8. Therefore each $f \in A$ with $\langle f,f\rangle = 0$ is an extreme point of A. Conversely, if $f \in A$ and $\langle f,f\rangle > 0$, then for sufficiently small ε we have $\langle f \pm \frac{\varepsilon}{z}, f \pm \frac{\varepsilon}{z}\rangle \geq 0$. This implies $f \pm \frac{\varepsilon}{z} \in A$ and, consequently, $f \notin E_A$. Thus E_A is precisely the set of functions $f \in A$ for which $\langle f,f\rangle = 0$.

<u>COROLLARY 8.10</u> (Springer [S17]). If $f \in \overline{co}\,\Sigma'$ and $\langle f,f\rangle = 0$, then $f \in E_{\overline{co}\,\Sigma'}$.

<u>Proof</u>. If $f \in \Sigma$, then the area theorem (Theorem B1; see also the example following Corollary 11.9) is just the inequality $\langle f,f\rangle \geq 0$. Therefore $\Sigma' \subset A$. Since A is closed and convex, $\overline{co}\,\Sigma' \subset A$. Extreme points of A that belong to $\overline{co}\,\Sigma'$ are necessarily extreme points of $\overline{co}\,\Sigma'$, so that the proof is complete.

If $f \in \Sigma$, then $\pi\langle f,f\rangle$ is the area of the complement of the range of f . Thus Corollary 8.10 says that each mapping in Σ' onto a domain whose complement has zero area is an extreme point of $\overline{co}\,\Sigma'$. This shows that $E_{\overline{co}\,\Sigma'}$ (as well as $E_{\Sigma'}$, since $E_{\overline{co}\,\Sigma'} \subset E_{\Sigma'}$ by the Krein-Milman theorem) is too large to be of value in studying linear extremal problems over Σ' .

<u>EXERCISE</u>. If $f \in \Sigma'$ and $C - f(|z| > 1)$ contains a nonempty open set, show that f is not an extreme point of Σ' .

<u>PROBLEM</u>. Are there extreme points of Σ' with $\langle f,f\rangle > 0$? What can be said between the conditions of Corollary 8.10 and the previous exercise?

CHAPTER 9. Elementary variational methods

In this chapter we consider the problem

$$\max_{\mathfrak{F}} \operatorname{Re} L$$

where $\mathfrak{F} = \mathfrak{F}(D,\ell_1,\ell_2,P,Q)$ is a nonempty compact family and $L \in H'(D)$. Since L is continuous and $\mathfrak{F}$ is compact, a maximum always occurs within the family $\mathfrak{F}$. By elementary variations we shall give some necessary conditions satisfied by extremal functions.

Suppose now $f \in \mathfrak{F}$ and $g = f + \varepsilon h + o(\varepsilon) \in H_u(D)$ where $\varepsilon > 0$ and $o(\varepsilon)/\varepsilon \to 0$ as $\varepsilon \to 0$ in the topology of $H(D)$, i.e. uniformly on compact subsets of D. Then

$$f^* = T(g) = [-1/\ell_o(g)][g - \tilde{\ell}_o(g)] = f + \varepsilon[h + \ell_o(h)f - \tilde{\ell}_o(h)] + o(\varepsilon)$$

is back in the family $\mathfrak{F}$. Consequently, if $L \in H'(D)$ and $\operatorname{Re} L(f) = \max_{\mathfrak{F}} \operatorname{Re} L$, then $\operatorname{Re} L(f^*) \le \operatorname{Re} L(f)$ or, equivalently,

$$\operatorname{Re}\{\varepsilon\, L[h + \ell_o(h)\, f - \tilde{\ell}_o(h)]\} + o(\varepsilon) \le 0 .$$

It is convenient to associate with L, f, and $\mathfrak{F}$ the new functional

$$L_f = L + L(f)\ell_o - L(1)\tilde{\ell}_o \in H'(D) .$$

Then the above condition is just

$$\operatorname{Re}\{\varepsilon\, L_f(h)\} + o(\varepsilon) \le 0 .$$

Since $\varepsilon > 0$, we may divide by ε and then let $\varepsilon \to 0$. The result is the inequality

$$\operatorname{Re} L_f(h) \le 0 .$$

We shall now introduce some elementary variations of the above form.

A. Rotation in U. If $f \in \mathfrak{F}(U, \ell_1, \ell_2, P, Q)$, then $g_{\pm}(z) = f(e^{\pm i\alpha}z) \in H_u(U)$ for all $\alpha > 0$. As $\alpha \to 0$,

$$g_{\pm} = f \pm i\alpha z f' + O(\alpha^2) .$$

Therefore, if $\Re L(f) = \max_{\mathfrak{F}} \Re L$, then

$$\Re\{\pm i L_f(zf')\} \leq 0$$

for both signs. Consequently, our first necessary condition for an extremal function is

$$\Im L_f(zf') = 0 .$$

B. Möbius self-mapping of U. If $f \in \mathfrak{F}(U, \ell_1, \ell_2, P, Q)$, $0 < r < 1$, and $\theta \in \mathbb{R}$, then $g(z) = f([z + re^{i\theta}]/[1 + re^{-i\theta}z]) \in H_u(U)$. As $r \to 0$,

$$g = f + r(e^{i\theta}f' - e^{-i\theta}z^2 f') + O(r^2) .$$

Therefore if $\Re L(f) = \max_{\mathfrak{F}} \Re L$, then

$$0 \geq \Re L_f(e^{i\theta}f' - e^{-i\theta}z^2 f') = \Re\{e^{i\theta}[L_f(f') - \overline{L_f(z^2 f')}]\}$$

for all $\theta \in \mathbb{R}$. Consequently, our second necessary condition for an extremal function is

$$L_f(f') = \overline{L_f(z^2 f')} .$$

C. Pre-slit-mapping in U. The Koebe function $k(z) = z/(1 + \eta z)^2$, $|\eta| = 1$, maps U onto the plane with a radial slit from $k(\bar{\eta}) = \frac{1}{4}\bar{\eta}$ to ∞. For $0 < \varepsilon < 1$ the function $s_\varepsilon(z) = k^{-1}((1-\varepsilon)k(z))$ maps U onto U with a slit from $\bar{\eta}$ omitted. If $f \in \mathfrak{F}(U, \ell_1, \ell_2, P, Q)$, then $g = f \circ s_\varepsilon \in H_u(U)$

and as $\epsilon \to 0$

$$g = f - \epsilon(zf')(1+\eta z)/(1-\eta z) + O(\epsilon^2) .$$

Therefore, if $\Re e\, L(f) = \max_{\mathfrak{F}} \Re e\, L$, then

$$\Re e\, L_f((zf')(1+\eta z)/(1-\eta z)) \geq 0 \quad \text{for all} \quad \eta, \ |\eta| = 1 .$$

Furthermore, if μ is any probability measure on $|\eta| = 1$, then by continuity of the functional L_f

$$\Re e\, L_f(zf' \int_{|\eta|=1} (1+\eta z)/(1-\eta z) d\mu) \geq 0 .$$

Since μ is arbitrary, it follows from Theorem 1.6 that

$$\Re e\, L_f(zf'p) \geq 0 \qquad \text{for all} \quad p \in P .$$

We summarize the above conditions in the following theorem.

THEOREM 9.1. Suppose $\mathfrak{F} = \mathfrak{F}(U, \ell_1, \ell_2, P, Q)$ is compact, $f \in \mathfrak{F}$, $L \in H'(U)$, and $\Re e\, L(f) = \max_{\mathfrak{F}} \Re e\, L$. If $L_f = L + L(f)\ell_o - L(1)\tilde{\ell}_o$, then

(A) $\Im m\, L_f(zf') = 0$

(B) $L_f(f') = L_f(z^2 f')$

(C) $\Re e\, L_f(zf'p) \geq 0$ for all $p \in P$.

REMARKS. For $p \equiv 1$ condition (C) complements (A) to say that $L_f(zf')$ is real and nonnegative. Also, by replacing L by $-L$, one finds the same conditions (A) and (B) in case $\Re e\, L(f) = \min_{\mathfrak{F}} \Re e\, L$.

EXAMPLE. Suppose $f \in \mathcal{S}(D, z_o)$ and $L \in H'(D)$. Since $\ell_o(g) = -g'(z_o)$ and $\tilde{\ell}_o(g) = g(z_o)$, the functional

$$L_f(g) = L(g) - g'(z_o)L(f) - g(z_o)L(1) .$$

For future use we note that

$$L_f(1/(f-w)) = (1/w^2)L(f^2/(f-w))$$

for each $w \in \mathbb{C}-f(D)$.

Secondly, suppose $f \in \mathfrak{T}(D,p,q,P,Q)$ and $L \in H'(D)$. Since $\ell_o(g) = [g(p)-g(q)]/(Q-P)$ and $\mathcal{L}_o(g) = [Qg(p)-Pg(q)]/(Q-P)$, the functional

$$L_f(g) = L(g) + [g(p)L(f-Q) - g(q)L(f-P)]/(Q-P) .$$

For future use we note that

$$L_f(1/(f-w)) = L\left(\frac{(f-P)(f-Q)}{(f-w)(P-w)(Q-w)}\right) .$$

<u>Application to the class S</u>. In the class S, the functional $L_f(g) = L(g) - g'(0)L(f) - g(0)L(1)$. Therefore, if f is an extremal function for the problem $\max_S \operatorname{Re} L$, then the conditions of Theorem 9.1 become

(A) $\operatorname{Im} L(zf'-f) = 0$

(B) $L(f'-1) - f''(0)L(f) = \overline{L(z^2f')}$

(C) $\operatorname{Re} L(f) \leq \operatorname{Re} L(zf'p)$ for all $p \in P$.

If we represent $L(f) = \sum_{n=0}^{\infty} a_n b_n$, where $f(z) = \sum_{n=0}^{\infty} a_n z^n$, as in Corollary 4.2, then the coefficients of an extremal function satisfy

(A) $\operatorname{Im} \sum_{n=2}^{\infty} (n-1)a_n b_n = 0$

(B) $\sum_{n=2}^{\infty} \{[(n+1)a_{n+1} - 2a_2 a_n]b_n - (n-1)\overline{a_{n-1}b_n}\} = 0$

(C) $\operatorname{Re} \left\{\sum_{n=2}^{\infty}\left[(n-1)a_n + c_{n-1} + \sum_{m=2}^{n-1} m a_m c_{n-m}\right]b_n\right\} \geq 0$

whenever $1 + \sum_{k=1}^{\infty} c_k z^k \in P$.

In particular, the coefficients of an extremal function for the problem $\max_S \operatorname{Re} a_n$, $n \geq 2$, satisfy

(A) $\operatorname{Im} a_n = 0$

(B) $(n+1)a_{n+1} - 2a_2 a_n - (n-1)\overline{a_{n-1}} = 0$ (Marty relations)

(C) $\operatorname{Re}\left\{(n-1)a_n + c_{n-1} + \sum_{m=2}^{n-1} m a_m c_{n-m}\right\} \geq 0$ whenever

$$1 + \sum_{k=1}^{\infty} c_k z^k \in P .$$

More generally, if f is an extremal function for the problem $\max_S \operatorname{Re} f^{(n)}(z_o)$ for fixed $z_o \in U$ and $n \geq 0$, then

(A) $\operatorname{Im}\{z_o f^{(n+1)}(z_o) + (n-1) f^{(n)}(z_o)\} = 0$

(B) $f'(z_o) - 1 - f''(0) f(z_o) = \overline{z_o^2 f'(z_o)}$ for $n = 0$,

$$f^{(n+1)}(z_o) - f''(0) f^{(n)}(z_o)$$

$$= \overline{z_o^2 f^{(n+1)}(z_o)} + \overline{2nz_o f^{(n)}(z_o)} + \overline{n(n-1) f^{(n-1)}(z_o)}$$

for $n \geq 1$,

(C) $\operatorname{Re}\{[1 - n(p(z_o) + z_o p'(z_o))] f^{(n)}(z_o)\}$

$$\leq \operatorname{Re}\{z_o p(z_o) f^{(n+1)}(z_o)$$

$$+ \sum_{k=1}^{n-1} [k\binom{n}{k} p^{(n-k)}(z_o) + \binom{n}{k-1} z_o p^{(n-k+1)}(z_o)] f^{(k)}(z_o)\}$$

for all $p \in P$.

In Appendix B we show that $\max_S |a_2| = 2$. Therefore the Koebe functions $k(z) = z/(1-\eta z)^2$, $\eta = \pm|b_2|/b_2$, are solutions to the problems $\max\limits_{S}\min \operatorname{Re} a_2 b_2$. Less elementary problems are $\max\limits_{S}\min \operatorname{Re}(a_2 b_2 + a_3 b_3)$.

EXERCISE. Show that the problems $\max\limits_{S}\min \operatorname{Re}(a_2 b_2 + a_3 b_3)$ can have Koebe functions $k(z) = z/(1-\eta z)^2$, $|\eta| = 1$, for solutions only if $(b_2)^2 \overline{b_3}$ is real.

<u>Solution</u>. If $k(z) = z/(1-\eta z)^2 = z + 2\eta z^2 + 3\eta^2 z^3 + 4\eta^3 z^4 + \ldots$ is a solution of $\max_S / \min_S \, \mathrm{Re}(a_2 b_2 + a_3 b_3)$, then by Theorem 9.1(A)

$$\mathrm{Im}(6\eta^2 b_3 + 2\eta b_2) = 0$$

and by Theorem 9.1(B)

$$4\eta^3 b_3 + \eta^2 b_2 - 4\overline{\eta b_3} - \overline{b_2} = 0 .$$

So $12\eta^2 b_3 + 3\eta b_2 + \overline{\eta b_2} = 12\overline{\eta}^2 \overline{b_3} + 4\overline{\eta b_2} = 12\eta^2 b_3 + 4\eta b_2$. By cancellation, ηb_2 is real, and hence, $\eta^2 b_3$ is real also. Therefore $(\eta b_2)^2 (\overline{\eta}^2 \overline{b_3}) = (b_2)^2 \overline{b_3}$ is real.

<u>PROBLEM</u>. Find $\max_S \mathrm{Re}(a_2 + ia_3)$. The Koebe function cannot be extremal by the previous exercise. The answer should be in the interval $(3\frac{1}{6}, 3\frac{1}{4})$. In theory the answer is known, for A. C. Schaeffer and D. C. Spencer [S1] have described the region of values of (a_2, a_3) in $\mathbb{C}^2$ as f varies over S .

<u>EXERCISE</u>. Show that the problems $\max_S / \min_S \, \mathrm{Re}\, f(z_o)$ can have Koebe functions $k(z) = z/(1-\eta z)^2$, $|\eta| = 1$, for solutions only if z_o is real and $\eta = \pm 1$. That is, Koebe functions never are extreme for $\mathrm{Re}\, f(z_o)$ if $z_o \notin \mathbb{R}$. Actually, using the region of values for $\log[f(z_o)/z_o]$ as f varies over S , namely, the disk $|w + \log(1-|z_o|^2)| \le \log[(1+|z_o|)/(1-|z_o|)]$, obtained by H. Grunsky [G10], one can show that a Koebe function is extremal for $\max_S \mathrm{Re}\, f(z_o)$ iff $\frac{1-e}{1+e} \le z_o < 1$ and $\eta = 1$.

<u>EXERCISE</u>. Consider the problem $\max_S \mathrm{Re}\, f(z_o)$. Make an optimal choice of $p(z) = (1+\eta z)/(1-\eta z)$ in Theorem 9.1(C) to show that an extremal function satisfies

$$\Re\, f(z_o) \le [(1+|z_o|^2)\, \Re\{z_o f'(z_o)\} - 2|z_o^2 f'(z_o)|]/[1-|z_o|^2] .$$

EXERCISE. On the basis of Theorem 9.1, give values of z_o such that Koebe functions are not solutions of the problem $\max_S \Re\, f^{(n)}(z_o)$.

We now return to arbitrary domains D. Except as $w \to \infty$, the following is not a variation in the usual sense, but it carries useful information.

D. Möbius post-mapping. Let $f \in \mathfrak{F}(D,\ell_1,\ell_2,P,Q)$ and $w \in C-f(D)$. Then $g = 1/(f-w) \in H_u(D)$ and $f^* = T(g) = [-1/\ell_o(1/(f-w))] \cdot [1/(f-w) - \tilde{\ell}_o(1/(f-w))] \in \mathfrak{F}$ as long as $\ell_o(1/(f-w)) \neq 0$. If $L \in H'(D)$ and $\Re\, L(f) = \max_{\mathfrak{F}} \Re\, L$, then the condition $\Re\, L(f^*) \le \Re\, L(f)$ becomes $\Re\{L_f(1/(f-w))/\ell_o(1/(f-w))\} \ge 0$ where $L_f = L + L(f)\ell_o - L(1)\tilde{\ell}_o$.

THEOREM 9.2. Suppose $\mathfrak{F} = \mathfrak{F}(D,\ell_1,\ell_2,P,Q)$ is compact, $f \in \mathfrak{F}$, $L \in H'(D)$, and $\Re\, L(f) = \max_{\mathfrak{F}} \Re\, L$. Then

$$\Re\{L_f(1/(f-w))/\ell_o(1/(f-w))\} \ge 0$$

for all $w \in C-f(D)$ such that $\ell_o(1/(f-w)) \neq 0$.

For later applications we shall need conditions which assure that $\ell_o(1/(f-w)) \neq 0$:

LEMMA 9.3. Suppose $\mathfrak{F} = \mathfrak{F}(D,\ell_1,\ell_2,P,Q)$ is compact, $f \in \mathfrak{F}$, and γ is a nondegenerate component of $C-f(D)$. Then

(a) either $\ell_o(1/(f-w)) \equiv 0$ on γ or $\ell_o(1/(f-w)) \neq 0$ for all $w \in \gamma$.

(b) If the support of some representing measure for ℓ_o does not separate γ from ∞, then $\ell_o(1/(f-w)) \neq 0$ for all $w \in \gamma$.

(c) If D has a strongly dense boundary, then $\ell_o(1/(f-w)) \neq 0$ for all $w \in \gamma$.

Proof. (a): Assume $\ell_o(1/(f-w_o)) = 0$ for some $w_o \in \gamma$, but $\ell_o(1/(f-w)) \not\equiv 0$ on γ. Then for some $\epsilon > 0$, $\ell_o(1/(f-w_o)) \neq 0$ for $0 < |w-w_o| < \varepsilon$. The inversion $t = 1/(w-w_o)$ maps γ onto a continuum Γ containing ∞, and $\ell_o(1/[f-(w_o + \frac{1}{t})]) \neq 0$ for $1/\varepsilon < |t| < \infty$. Let $g = 1/(f-w_o)$. Then $g \in H_u(D)$ and $\ell_o(g) = 0$. For $t_1 \in \Gamma \cap \{1/\varepsilon < |t| < \infty\}$, the expression

$$\ell_o(1/(g-t_1)) = \frac{-1}{t_1}\ell_o(1) - \frac{1}{t_1^2}\ell_o(1/[f-(w_o+\frac{1}{t_1})]) = \frac{-1}{t_1^2}\ell_o(1/[f-(w_o+\frac{1}{t_1})])$$

is nonzero. Therefore, based on the function g we may construct variations $g_n = g + \epsilon_n/(g-t_1) + o(\epsilon_n)$ such that $f_n = T(g_n) \in \mathfrak{F}$ and the sequence $\{f_n\}$ contradicts the compactness of $\mathfrak{F}$, just as in the proof of Theorem 7.3(b).

(b): If $\ell_o(1/(f-w)) \equiv 0$ on γ, then the analytic function $\ell_o(1/(f-w)) \equiv 0$ in a neighborhood of ∞ also, since the support of some representing measure for ℓ_o does not separate γ from ∞. However, near ∞

$$\ell_o(1/(f-w)) = -\frac{1}{w}\ell_o(1) - \frac{1}{w^2}\ell_o(f) - \frac{1}{w^3}\ell_o(f^2/(1-\frac{1}{w})) = \frac{1}{w^2} + o(\frac{1}{w^3}),$$

so that ∞ is a zero of order 2.

(c): If D has a strongly dense boundary, then $\ell_o(1/(f-w)) \neq 0$ for all $w \in \mathbb{C}-f(D)$ by Theorem 7.3(b), since $1/(f-w) \in H_u(D)$.

It is also useful to know conditions under which $L_f(1/(f-w)) \neq 0$.

LEMMA 9.4. Suppose $\mathfrak{F} = \mathfrak{F}(D,\ell_1,\ell_2,P,Q)$ is compact, $f \in \mathfrak{F}$, $L \in H'(D)$ is linearly independent of ℓ_1 and ℓ_2, and the support of some representing measure for $L_f = L + L(f)\ell_o - L(1)\tilde{\ell}_o$ does not separate the components of $\bar{C}-D$. Then

(a) $L_f(1/(f-w)) \not\equiv 0$ in a neighborhood of ∞, and

(b) $L_f(1/(f-w)) \not\equiv 0$ on each nondegenerate component of $C-f(D)$.

Proof. Let K be the support of a representing measure for L_f, that does not separate the components of $\bar{C}-D$. Then only the unbounded component O of $\bar{C}-K$ contains points of $\bar{C}-D$, and $E = C-O$ is a compact subset of D containing K. The function $L_f(1/(f-w))$ is analytic in the domain $C-f(E)$. If it vanishes identically in a neighborhood of ∞ or on a nondegenerate component of $C-f(D)$, then it vanishes identically on $C-f(E)$ and by Lemma 4.5, $L_f \equiv 0$ on $H(D)$. In this case L is a linear combination of ℓ_o and $\tilde{\ell}_o$, hence of ℓ_1 and ℓ_2.

In order to obtain some geometric information from Theorem 9.2 we shall use the following elementary fact:

LEMMA 9.5. Suppose that $\varphi(w) = \sum_{n=1}^{\infty} A_n w^{-n}$ is analytic in a neighborhood of ∞. Let $\{\theta_j\}$ be dense in $[0,2\pi]$, and assume that $\Re e\, \varphi(w_{jk}) \geq 0$ for a sequence $\{w_{jk}\}$ such that $\lim_{k\to\infty} w_{jk} = \infty$ and $\lim_{k\to\infty} w_{jk}/|w_{jk}| = e^{i\theta_j}$ for each j. Then $\varphi \equiv 0$.

Proof. Assume $\varphi \not\equiv 0$ and let A_N be the first nonzero coefficient. Then

$$0 \leq \lim_{k\to\infty} |w_{jk}|^N \Re e\, \varphi(w_{jk}) = \lim_{k\to\infty} \Re e \sum_{n=N}^{\infty} A_n |w_{jk}|^{N-n} (w_{jk}/|w_{jk}|)^{-n} = \Re e\, A_N e^{-iN\theta_j}$$

for a dense set $\{\theta_j\}$ in $[0,2\pi]$. Therefore $A_N = 0$, and we have a contradiction.

DEFINITION. A real number α is a limiting direction of a set A at ∞ if there exist points $w_n \in A$ with $\lim_{n\to\infty} w_n = \infty$ and $\lim_{n\to\infty}(w_n/|w_n|) = e^{i\alpha}$.

Clearly, any unbounded set has at least one limiting direction at ∞. The following theorem says that extremal functions for certain problems have the property that $\mathbb{C}-f(D)$ cannot "wind" infinitely often around ∞.

THEOREM 9.6. Suppose $\mathfrak{F} = \mathfrak{F}(D,\ell_1,\ell_2,P,Q)$ is compact, $f \in \mathfrak{F}$, $L \in H'(D)$ is linearly independent of ℓ_1 and ℓ_2, $\operatorname{Re} L(f) = \max_{\mathfrak{F}} \operatorname{Re} L$, and the support of some representing measure for L_f does not separate the components of $\bar{\mathbb{C}}-D$. Then $\mathbb{C}-f(D)$ cannot have a dense set of limiting directions at ∞.

Proof. Since $\ell_o(1) = 0$, $\tilde{\ell}_o(1) = 1$, $L_f(1) = 0$, $\ell_o(f) = -1$, $\tilde{\ell}_o(f) = 0$, $L_f(f) = 0$, and $1/(f-w) = -\frac{1}{w} - \frac{1}{w^2}f - \frac{1}{w^3}f^2/(1-\frac{1}{w}f)$, the function $\ell_o(1/(f-w))$ has a zero of order 2 at ∞ and $L_f(1/(f-w))$ has a zero of order at least 3 at ∞. Therefore $\varphi(w) = L_f(1/(f-w))/\ell_o(1/(f-w))$ is analytic in a neighborhood of ∞ and vanishes at ∞. If $\mathbb{C}-f(D)$ has a dense set of limiting directions at ∞, then by Theorem 9.2 we have $\operatorname{Re}\varphi(w_{jk}) \geq 0$ for a sequence $\{w_{jk}\}$ as in the hypothesis of Lemma 9.5. Therefore $\varphi \equiv 0$ in a neighborhood of ∞. This implies $L_f(1/(f-w)) \equiv 0$ in a neighborhood of ∞ and contradicts Lemma 9.4(a).

EXERCISE. Formulate other versions of Lemma 9.5 (e.g., with $\{\theta_j\}$ dense in $[\beta,\beta+\pi+\varepsilon]$ or with $\theta_j = \beta + (\pi+\varepsilon)/j$, $j=1,2,\ldots$, for some $\beta \in \mathbb{R}$ and $\varepsilon > 0$), and deduce analogous versions of Theorem 9.6 .

E. Variations relative to an exterior point. We shall construct a variation that will be useful when $\mathbb{C}-f(D)$ contains a non-empty open set.

LEMMA 9.7. Let $\tilde{D} = \{w: |w-w_o| > \rho\}$ and $\alpha \in \mathbb{R}$. Then the function $w + \rho^2 e^{2i\alpha}/(w-w_o)$ is univalent in $\tilde{D}$ and maps $\tilde{D}$ onto the complement of the line segment of length 4ρ and inclination α, centered at w_o.

Proof. $z(w) = e^{-i\alpha}(w-w_o)/\rho$ maps $\tilde{D}$ onto $|z|>1$; $t(z) = z + \frac{1}{z}$ maps $|z|>1$ onto the complement of the real interval $[-2,2]$; $\zeta(t) = w_o + \rho e^{i\alpha}t$ maps the complement of $[-2,2]$ onto the desired domain; and $(\zeta \circ t \circ z)(w) = w + \rho^2 e^{2i\alpha}/(w-w_o)$.

Suppose now $f \in \mathfrak{F}(D,\ell_1,\ell_2,P,Q)$ and $\mathbb{C}-f(D)$ contains an open set O. If $w_o \in O$, then by Lemma 9.7

$$g = f + \rho^2 e^{2i\alpha}/(f-w_o) \in H_u(D)$$

for all sufficiently small $\rho > 0$ and all $\alpha \in \mathbb{R}$. If $\operatorname{Re} L(f) = \max_{\mathfrak{F}} \operatorname{Re} L$, then just as with the earlier variations

$$\operatorname{Re}\{e^{2i\alpha} L_f(1/(f-w_o))\} \le 0 \text{ for every } \alpha \in \mathbb{R}.$$

Consequently, $L_f(1/(f-w_o)) = 0$. We have proved the following:

THEOREM 9.8. Suppose $\mathfrak{F} = \mathfrak{F}(D,\ell_1,\ell_2,P,Q)$ is compact, $f \in \mathfrak{F}$,

$L \in H'(D)$, and $\operatorname{Re} L(f) = \max_{\mathfrak{F}} \operatorname{Re} L$. If $\mathbb{C}-f(D)$ contains an open set O , then $L_f(1/(f-w)) \equiv 0$ in O , where $L_f = L + L(f)\ell_o - L(1)\tilde{\ell}_o$.

This has the following important consequence.

THEOREM 9.9. Suppose $\mathfrak{F} = \mathfrak{F}(D,\ell_1,\ell_2,P,Q)$ is compact, $f \in \mathfrak{F}$, $L \in H'(D)$ is linearly independent of ℓ_1 and ℓ_2 , $\operatorname{Re} L(f) = \max_{\mathfrak{F}} \operatorname{Re} L$, and the support of some representing measure for L_f does not separate the components of $\bar{\mathbb{C}}-D$. Then $\mathbb{C}-f(D)$ contains no nonempty open set.

Proof. If $\mathbb{C}-f(D)$ contains a nonempty open set O, then $L_f(1/(f-w)) \equiv 0$ in O by Theorem 9.8. Since the support K of some representing measure for L_f does not separate the components of $\tilde{\mathbb{C}}-D$, we have $L_f(1/(f-w)) \equiv 0$ on $\mathbb{C}-f(K)$, contradicting Lemma 9.4.

We now observe that the elementary methods of this chapter apply also to many nonlinear problems.

DEFINITION. Let λ be a real functional on $\mathfrak{F} \subset H(D)$. We shall say that λ has a complex Gâteaux derivative at $f \in \mathfrak{F}$ relative to $\mathfrak{F}$ if there exists an $L \in H'(D)$ (depending on f) such that

$$\lambda(f^*) = \lambda(f) + \varepsilon \operatorname{Re} L(h) + o(\varepsilon)$$

whenever $f^* \in \mathfrak{F}$, $\varepsilon > 0$, and $f^* = f + \varepsilon h + o(\varepsilon)$. The latter $o(\varepsilon)$ terms are measured in the topology of $H(D)$, i.e., $o(\varepsilon)/\varepsilon \to 0$ uniformly on compact subsets of D as $\varepsilon \to 0$.

THEOREM 9.10. Suppose $\mathfrak{F} = \mathfrak{F}(D,\ell_1,\ell_2,P,Q)$ is compact, $f \in \mathfrak{F}$, λ is a real continuous functional on $\mathfrak{F}$ with complex Gâteaux derivative $L \in H'(D)$ at f relative to $\mathfrak{F}$, $L_f = L + L(f)\ell_o - L(1)\tilde{\ell}_o$, and $\lambda(f) = \max_{\mathfrak{F}} \lambda$. If $D = U$, then

(A) $\mathscr{I}m\, L_f(zf') = 0$

(B) $L_f(f') = L_f(z^2 f')$

(C) $\mathscr{R}e\, L_f(zf'p) \geq 0$ for all $p \in P$.

Moreover, if D is arbitrary, L is linearly independent of ℓ_1, ℓ_2, and the support of some representing measure for L_f does not separate the components of $\bar{C}-D$, then $C-f(D)$ contains no non-empty open set.

Proof. In Theorems 9.1 and 9.8 (hence 9.9) additional terms of order $o(\varepsilon)$ are insignificant.

REMARK. If D is simply connected, then the conditions in Lemmas 9.3, 9.4, and Theorems 9.6, 9.9, 9.10, concerning supports of representing measures for certain linear functionals, are trivially satisfied.

CHAPTER 10. Application of Schiffer's boundary variation to linear problems

The elementary variations of Chapter 9 basically give implicit restrictions for extremal functions. However, alone they are not powerful enough to lead to solutions of substantial problems. A more useful variation (Schiffer's boundary variation) and a fine analysis of its implications (Schiffer's fundamental lemma) are contained in Appendix C. We shall first be concerned with consequences for linear problems.

THEOREM 10.1. Suppose $\mathfrak{F} = \mathfrak{F}(D, \ell_1, \ell_2, P, Q)$ is compact, $f \in \mathfrak{F}$, $L \in H'(D)$, $\operatorname{Re} L(f) = \max_{\mathfrak{F}} \operatorname{Re} L$, $L_f = L + L(f)\ell_o - L(1)\tilde{\ell}_o$, and γ is a nondegenerate component of $\mathbb{C}-f(D)$. If $L_f(1/(f-w)) \not\equiv 0$ on γ, then γ consists of finitely many analytic arcs each satisfying

$$L_f(1/(f-w))(dw)^2 > 0 .$$

The only possible points of nonanalyticity or branching of γ are the zeros of $L_f(1/(f-w))$. Consequently, if $L_f(1/(f-w))$ does not vanish on γ, then γ is a single analytic arc satisfying $L_f(1/(f-w))(dw)^2 > 0$.

Proof. The function $L_f(1/(f-w))$ is analytic off of a compact set in the complement of γ. If $L_f(1/(f-w)) \not\equiv 0$ on γ, then $L_f(1/(f-w))$ has at most finitely many zeros on γ.

Let γ_o be an arbitrary bounded subcontinuum of γ on which $L_f(1/(f-w)) \neq 0$. If $w_o \in \gamma_o$ and F_ν are as in the hypotheses of Theorem C4, then $g_\nu = F_\nu \circ f \in H_u(D)$. Just as at the beginning of Chapter 9, $f_\nu^* = T(g_\nu) \in \mathfrak{F}$ for all ν sufficiently large, and

$$\Re\{\rho_\nu^2 B_{1,\nu} L_f(1/(f-w_o))\} + O(\rho_\nu^3) \le 0 .$$

Dividing by ρ_ν^2 and letting $\nu \to \infty$, we have

$$\Re\{cL_f(1/(f-w_o))\} \le 0 .$$

We conclude from Theorem C4 that γ_o is an analytic arc satisfying

$$L_f(1/(f-w))(dw)^2 > 0 .$$

Since γ_o is arbitrary, the only possible points of non-analyticity or branching of γ are the zeros of $L_f(1/(f-w))$. At a zero of $L_f(1/(f-w))$, an analysis of the differential equation $L_f(1/(f-w))(dw)^2 > 0$ shows that only finite branching is possible (cf. J. A. Jenkins [J1, Chapter III]). Since there are at most finitely many zeros, we conclude that γ consists of finitely many analytic arcs.

<u>REMARK</u>. $L_f(1/(f-w))(dw)^2 > 0$ is a functional differential equation in the sense that it depends on the extremal function f .

Before stopping for examples, we shall obtain some important general properties of extremal functions to linear extremal problems.

<u>LEMMA 10.2</u>. Suppose $\mathfrak{F} = \mathfrak{F}(D,\ell_1,\ell_2,P,Q)$ is compact, $f \in \mathfrak{F}$, $L \in H'(D)$, and $\Re L(f) = \max_{\mathfrak{F}} \Re L$. If γ is a nondegenerate component of $\mathbb{C}-f(D)$ on which $L_f(1/(f-w)) \not\equiv 0$ and $\ell_o(1/(f-w)) \not\equiv 0$, then γ is a single analytic arc. Moreover, if $L_f(1/(f-w))$ has a zero $w_o \in \gamma$, then γ lies on the straight line $\{w_o + t/\sqrt{\ell_o(1/(f-w_o))} : t \in (-\infty,\infty)\}$.

<u>Proof</u>. Let $\mathfrak{F}$, f, L, γ, L_f, and ℓ_o be as in the hypotheses.

From Theorem 10.1 the only possible points of nonanalyticity or branching of γ are the zeros of $L_f(1/(f-w))$. However, if $L_f(1/(f-w))$ has a zero on γ, we shall show that γ lies on a straight line, hence is an analytic arc.

Assume therefore that $L_f(1/(f-w_o)) = 0$ for some $w_o \in \gamma$. Then $L_f(1/(f-w)) \neq 0$ for all $w \neq w_o$ in a neighborhood of w_o, since $L_f(1/(f-w)) \not\equiv 0$ on γ. Furthermore, $\ell_o(1/(f-w))$ never vanishes on γ by Lemma 9.3(a). Since $w_o \in \gamma$, the function $1/(f-w_o) \in H_u(D)$ and

$$\hat{f} = T(1/(f-w_o)) = [-1/\ell_o(1/(f-w_o))][1/(f-w_o) - \tilde{\ell}_o(1/(f-w_o))]$$

belongs to $\mathfrak{F}$. At the same time, the mapping

$$\hat{w} = \hat{w}(w) = [-1/\ell_o(1/(f-w_o))][1/(w-w_o) - \tilde{\ell}_o(1/(f-w_o))]$$

takes γ onto a continuum $\hat{\gamma} \subset \bar{\mathbb{C}}-f(D)$ containing $\infty = \hat{w}(w_o)$. We note for future use that

$$\frac{d\hat{w}}{dw} = 1/[\ell_o(1/(f-w_o))(w-w_o)^2] .$$

Observe now that

$$L(\hat{f}) = [-1/\ell_o(1/(f-w_o))][L(1/(f-w_o)) - \tilde{\ell}_o(1/(f-w_o))L(1)]$$

$$= [-1/\ell_o(1/(f-w_o))][L_f(1/(f-w_o)) - L(f)\ell_o(1/(f-w_o))] = L(f),$$

so that $\hat{f}$ is also an extremal function for the problem $\max_{\mathfrak{F}} \operatorname{Re} L$. Since $L(\hat{f}) = L(f)$, the functionals $L_{\hat{f}} = L_f$, and by direct computation

$$L_{\hat{f}}(1/(\hat{f}-\hat{w})) = L_f(1/(\hat{f}-\hat{w})) = (w-w_o)\ell_o(1/(f-w_o))L_f((f-w_o)/(f-w))$$

$$= (w-w_o)^2\ell_o(1/(f-w_o))L_f(1/(f-w))$$

since $L_f(1) = 0$. Evidently, $L_{\hat{f}}(1/(\hat{f}-\hat{w})) \not\equiv 0$ on $\hat{\gamma}$. Since $\hat{f}$ is also an extremal function, we may apply Theorem 10.1, this time to $\hat{\gamma}$, to learn that $\hat{\gamma}$ consists of analytic arcs satisfying the differential equation

$$L_{\hat{f}}(1/(\hat{f}-\hat{w}))(d\hat{w})^2 > 0$$

For all $w \in \gamma$, $w \neq w_o$, in a sufficiently small neighborhood of w_o the quotient

$$\frac{L_{\hat{f}}(1/(\hat{f}-\hat{w}))(d\hat{w})^2}{L_f(1/(f-w))(dw)^2} = (w-w_o)^2 \ell_o(1/(f-w_o))\left(\frac{d\hat{w}}{dw}\right)^2 = 1/[(w-w_o)^2 \ell_o(1/(f-w_o))]$$

is positive. That is, w must lie on the straight line $\{w_o + t\sqrt{\ell_o(1/(f-w_o))} : t \in (-\infty,\infty)\}$. In particular, γ is an analytic arc in a neighborhood of w_o.

We have shown in any case that γ is a single analytic arc. Furthermore, if $L_f(1/(f-w))$ has a zero on γ, then γ lies locally, hence globally by its analyticity, on the indicated line.

<u>LEMMA 10.3</u>. Suppose $\mathfrak{F} = \mathfrak{F}(D,\ell_1,\ell_2,P,Q)$ is compact, $f \in \mathfrak{F}$, $L \in H'(D)$, and $\operatorname{Re} L(f) = \max_{\mathfrak{F}} \operatorname{Re} L$. If γ is a nondegenerate component of $\mathbb{C}-f(D)$ on which $L_f(1/(f-w)) \not\equiv 0$ and $\ell_o(1/(f-w)) \not\equiv 0$, then γ is an analytic arc whose tangent makes an angle of at most $\pi/4$ with respect to the vector field $\operatorname{grad}[\operatorname{Re}\int\sqrt{\ell_o(1/(f-w))}\,dw]$.

<u>Proof</u>. The analyticity of γ is a consequence of Lemma 10.2. Fix $w_o \in \gamma$. If $L_f(1/(f-w_o)) = 0$, then by Lemma 10.2, γ lies on a line that has the same direction as the vector field $\operatorname{grad}[\operatorname{Re}\int\sqrt{\ell_o(1/(f-w))}\,dw]$ at w_o. If $L_f(1/(f-w_o)) \neq 0$, then we use the condition

$$\Re e\{L_f(1/(f-w_o))/\ell_o(1/(f-w_o))\} \geq 0$$

of Theorem 9.2. Note that $\ell_o(1/(f-w_o)) \neq 0$ by Lemma 9.3(a). In addition, $L_f(1/(f-w))(dw)^2 > 0$ at w_o by Theorem 10.1. Therefore the quotient

$$\Re e\{1/[\ell_o(1/(f-w_o))(dw)^2]\} \geq 0$$

at w_o; hence $|\arg[\sqrt{\ell_o(1/(f-w_o))}dw]^2| \leq \pi/2$ at w_o. We choose first a branch of $\sqrt{\ell_o(1/(f-w))}$ on γ and then the tangent direction so that

$$|\arg[\sqrt{\ell_o(1/(f-w_o))}dw]| \leq \pi/4$$

at w_o. The conclusion is then immediate.

The following theorem summarizes some analytic and geometric properties of solutions to linear extremal problems.

<u>THEOREM 10.4</u> ($\pi/4$ - theorem; Hengartner and Schober [H5]). Suppose $\mathfrak{F} = \mathfrak{F}(D,\ell_1,\ell_2,P,Q)$ is compact, $f \in \mathfrak{F}$, $L \in H'(D)$ is linearly independent of ℓ_1, ℓ_2, and $\Re e\, L(f) = \max_{\mathfrak{F}} \Re e\, L$. Assume furthermore that the supports of some representing measures for ℓ_o and L_f do not separate the components of $\bar{\mathbb{C}}-D$. Then each non-degenerate component γ of $\mathbb{C}-f(D)$ is a single analytic arc whose tangent makes an angle of at most $\pi/4$ with the vector field $\operatorname{grad}[\Re e\int\sqrt{\ell_o(1/(f-w))}dw]$. If $L_f(1/(f-w))$ vanishes at a point $w_o \in \gamma$, then γ lies on the straight line $\{w_o + t/\sqrt{\ell_o(1/(f-w_o))} : t \in (-\infty,\infty)\}$. Otherwise, γ satisfies $L_f(1/(f-w))(dw)^2 > 0$. At most one component of $\mathbb{C}-f(D)$ is unbounded.

Proof. Except for the final assertion, Theorem 10.4 follows from Theorem 10.1 and Lemmas 10.2 and 10.3 by inserting conditions from Lemmas 9.3(b) and 9.4(b) which guarantee that $L_f(1/(f-w)) \neq 0$ and $\ell_o(1/(f-w)) \neq 0$ on each nondegenerate component of $\mathbb{C}-f(D)$. To see that at most one component of $\mathbb{C}-f(D)$ is unbounded, we use an idea of L. Brickman and D. R. Wilken [B11]. If indeed two components of $\mathbb{C}-f(D)$ were unbounded, they would belong to a single component of $\bar{\mathbb{C}}-f(D)$. This component would then contain at least two distinct points on each Jordan curve that winds around ∞ in a given neighborhood of ∞. By Lemmas 8.1 and 8.2 we then have a decomposition $f = \lambda f_1 + (1-\lambda) f_2$ where $\lambda \in (0,1)$ and $f_1, f_2 \in \mathfrak{F}$, but both f_1 and f_2 omit nonempty open sets. Since this is a convex decomposition, both f_1 and f_2 also maximize $\operatorname{Re} L$ over $\mathfrak{F}$. However, f_1 and f_2 contradict the first assertion, that extremal functions must map onto the complement of (analytic) arcs and points.

A notable special case is the following:

COROLLARY 10.5. Suppose $D \neq \mathbb{C}$ is simply connected, $\mathfrak{F} = \mathfrak{F}(D,\ell_1,\ell_2,P,Q)$ is compact, $f \in \mathfrak{F}$, $L \in H'(D)$ is linearly independent of ℓ_1, ℓ_2, and $\operatorname{Re} L(f) = \max_{\mathfrak{F}} \operatorname{Re} L$. Then $\mathbb{C}-f(D)$ is a single analytic arc whose tangent makes an angle of at most $\pi/4$ with the vector field $\operatorname{grad}[\operatorname{Re}\int\sqrt{\ell_o(1/(f-w))}\,dw]$. If $L_f(1/(f-w))$ vanishes at a point $w_o \in \mathbb{C}-f(D)$, then $\mathbb{C}-f(D)$ lies on the straight line $\{w_o + t/\sqrt{\ell_o(1/(f-w_o))} : t \in (-\infty,\infty)\}$; otherwise, $\mathbb{C}-f(D)$ satisfies $L_f(1/(f-w))(dw)^2 > 0$.

We now apply Theorem 10.4 to the important families $\mathcal{S}(D,z_o)$ and $\mathcal{T}(D,p,q,P,Q)$. In complex coordinates $\mathrm{grad}[\mathrm{Re}\int\sqrt{\ell_o(1/(f-w))}\,dw] = \frac{1}{2}\overline{\sqrt{\ell_o(1/(f-w))}}$. Therefore the direction of the field is the argument of $1/\sqrt{\ell_o(1/(f-w))}$.

For $\mathcal{S}(D,z_o)$, one has $\ell_o(1/(f-w)) = 1/w^2$. Therefore the trajectories of $\mathrm{grad}[\mathrm{Re}\int(1/w)\,dw]$ are the rays from the origin.

For $\mathcal{T}(D,p,q,P,Q)$, one has $\ell_o(1/(f-w)) = 1/[(w-P)(w-Q)]$. Therefore the direction of the vector field at each point is the same as the argument of $\sqrt{\overline{(w-P)(w-Q)}}$. This direction is normal to the ellipse through w with foci P and Q. Therefore the trajectories of the vector field $\mathrm{grad}[\mathrm{Re}\int 1/\sqrt{(w-P)(w-Q)}\,dw]$ are the hyperbolae with foci P and Q.

We shall also use the expressions for $L_f(1/(f-w))$ from the example on pp. 81-82.

COROLLARY 10.6. Suppose $f \in \mathcal{S}(D,z_o)$, $L \in H'(D)$ is not of the form $L(g) = \alpha g(z_o) + \beta g'(z_o)$, $\mathrm{Re}\, L(f) = \max_{\mathcal{S}} \mathrm{Re}\, L$, and the support of some representing measure for L does not separate the components of $\bar{\mathbb{C}}-D$. Then each nondegenerate component γ of $\mathbb{C}-f(D)$ is a single analytic arc whose tangent makes an angle of at most $\pi/4$ with the radial direction. If $L(f^2/(f-w))$ vanishes at a point $w_o \in \gamma$, then γ lies on a ray from the origin through w_o; otherwise, γ satisfies $L(f^2/(f-w))(\frac{dw}{w})^2 > 0$. At most one component of $\mathbb{C}-f(D)$ is unbounded.

COROLLARY 10.7. Suppose $f \in \mathcal{T}(D,p,q,P,Q)$, $L \in H'(D)$ is not of the form $L(g) = \alpha g(p) + \beta g(q)$, $\mathrm{Re}\, L(f) = \max_{\mathcal{T}} \mathrm{Re}\, L$, and the support of some representing measure for L does not separate the components of $\bar{C}-D$. Then each nondegenerate component γ of $C-f(D)$ is an analytic arc whose tangent makes an angle of at most $\pi/4$ with respect to the family of hyperbolae with foci P and Q. If $L\left(\frac{(f-P)(f-Q)}{(f-w)(P-w)(Q-w)}\right)$ vanishes at a point $w_o \in \gamma$, then γ lies on a line that is tangent at w_o to the hyperbola through w_o with foci P and Q; otherwise, γ satisfies $L\left(\frac{(f-P)(f-Q)}{(f-w)(P-w)(Q-w)}\right)(dw)^2 > 0$. At most one component of $C-f(D)$ is unbounded.

Corollary 10.7 is a geometrically pleasing complement to Corollary 10.6 since, as $P,Q \to 0$, the hyperbolae degenerate into rays and the corresponding differentials coincide.

<u>DEFINITION</u>. f is a <u>support point</u> of a family $\mathfrak{F} \subset H(D)$ if $f \in \mathfrak{F}$ and there exists an $L \in H'(D)$ that is nonconstant on $\mathfrak{F}$, such that $\Re e\, L(f) = \max_{\mathfrak{F}} \Re e\, L$.

Geometrically, at a support point the family has a supporting hyperplane. For simply connected domains we may phrase Corollaries 10.5, 10.6, and 10.7 in terms of support points:

<u>COROLLARY 10.8</u>. Suppose $D \neq C$ is simply connected, $\mathfrak{F} = \mathfrak{F}(D,\ell_1,\ell_2,P,Q)$ is compact, and f is a support point of $\mathfrak{F}$. Then $C-f(D)$ is a single analytic arc whose tangent makes an angle of at most $\pi/4$ with the vector field $\mathrm{grad}[\Re e\int\sqrt{\ell_o}(1/(f-w))dw]$.

If $\mathfrak{F} = \mathfrak{S}(D,z_o)$, the arc $C-f(D)$ makes an angle of at most $\pi/4$ with the radial direction.

If $\mathfrak{F} = \mathfrak{T}(D,p,q,P,Q)$, the arc $C-f(D)$ makes an angle of at most $\pi/4$ with the family of hyperbolae with foci P and Q.

If an arc continually makes an angle of at most $\pi/4$ with the radial direction, then it is monotone in the sense of Theorem 8.5. For the special case of the class S we therefore have the following consequence:

COROLLARY 10.9. The conclusions of Theorem 8.5 apply to the support points of S.

REMARKS. For the class S, a special case of Corollary 10.8 is contained in G. M. Goluzin's book [G8, p. 147] and the general case was proved by A. Pfluger [P4] and L. Brickman and D. R. Wilken [B11]. That $\bar{\mathbb{C}}-f(U)$ does not branch at ∞ for the coefficient problem in the class S was first proved by A. C. Schaeffer and D. C. Spencer [S1]. That there are no finite points of nonanalyticity for the same problem was first proved by M. Schiffer [S3].

Let us now consider the differential equation of Theorem 10.1. In case $D = U$, we may parametrize $\partial f(U)$ by $w = f(e^{i\theta})$. Then $\frac{dw}{d\theta} = i\zeta f'(\zeta)$, $\zeta = e^{i\theta}$, and the differential equation becomes

$$L_f(\zeta^2[f'(\zeta)]^2/[f(\zeta)-f(z)]) > 0$$

where $|\zeta| = 1$ and the linear functional is applied to the function of z. We now wish to extend this relation to $|\zeta| < 1$.

THEOREM 10.10. Suppose $\mathfrak{F} = \mathfrak{F}(U,\ell_1,\ell_2,P,Q)$ is compact, $f \in \mathfrak{F}$, $L \in H'(U)$ is linearly independent of ℓ_1, ℓ_2, and $\operatorname{Re} L(f) = \max_{\mathfrak{F}} \operatorname{Re} L$. Then f satisfies the functional differential equation

$$L_f\left(\frac{\zeta[f'(\zeta)]^2}{f(\zeta)-f(z)} + \frac{zf'(z)}{z-\zeta}\right) = L_f\left(\frac{z^2f'(z)}{1-\bar{\zeta}z}\right) \quad \text{for} \quad |\zeta| \le 1$$

where $L_f = L + L(f)\ell_o - L(1)\tilde{\ell}_o$.

<u>Proof</u>. By Theorem 10.1 and Corollary 10.5, $\mathbb{C}-f(U)$ is a single analytic arc satisfying the differential equation $L_f(1/(f-w))(dw)^2 \ge 0$, which is equivalent to the statement that $\sqrt{L_f(1/(f-w))}dw$ is real. Consequently,

$$\int_\infty^w \sqrt{L_f(1/(f-w))}\,dw$$

is real if the path is restricted to $\mathbb{C}-f(U)$. Since $L_f(1) = L_f(f) = 0$,

$$L_f(1/(f-w)) = -\frac{1}{w^3} L_f(f^2/(1-\frac{f}{w}))$$

so that the integral converges at the lower limit. We parametrize $\bar{\mathbb{C}}-f(U)$ by $f(\zeta)$, $|\zeta| = 1$. Then the (possibly multivalued) function

$$F(\zeta) = \int_\infty^{f(\zeta)} \sqrt{L_f(1/(f-w))}\,dw$$

is real, finite, and continuous on $|\zeta| = 1$. For some $r < 1$, the support of some representing measure for L_f is in $|z| \le r$. Therefore $F(\zeta)$ has an analytic extension to $r < |\zeta| < 1$. By the Schwarz reflection principle F has an analytic continuation across each point of $|\zeta| = 1$. Therefore

$$[F'(\zeta)]^2 = L_f([f'(\zeta)]^2/[f(z)-f(\zeta)])$$

is analytic in $r < |\zeta| < 1$ and has a finite analytic continuation across $|\zeta| = 1$.

It follows from Theorem 9.1(A) that $L_f(zf')$ is real. So for $|\zeta| = 1$

$$L_f\left(\frac{\zeta zf'(z)}{z-\zeta}\right) - \overline{L_f\left(\frac{\bar{\zeta}z^2f'(z)}{1-\bar{\zeta}z}\right)}$$

is real. Adding $L_f(\zeta^2[f'(\zeta)]^2/[f(\zeta)-f(z)])$, which is nonnegative for $|\zeta| = 1$, we find that

$$G(\zeta) = L_f\left(\frac{[\zeta f'(\zeta)]^2}{f(\zeta)-f(z)} + \frac{\zeta z f'(z)}{z-\zeta}\right) - \overline{L_f\left(\frac{\bar{\zeta} z^2 f'(z)}{1-\bar{\zeta} z}\right)}$$

is real on $|\zeta| = 1$. In fact,

$$\frac{[\zeta f'(\zeta)]^2}{f(\zeta)-f(z)} + \frac{\zeta z f'(z)}{z-\zeta}$$

has a removable singularity at $\zeta = z$. Therefore $G(\zeta)$ is defined and analytic for $|\zeta| \leq 1$ and is real on $|\zeta| = 1$. By the Schwarz reflection principle G extends to a bounded analytic function in $\mathbb{C}$. By Liouville's theorem $G(\zeta) \equiv G(0) = 0$, from which the theorem follows.

<u>REMARKS</u>. We actually proved that the identity of Theorem 10.10 holds in a neighborhood of $|\zeta| \leq 1$. This does not mean that $f'(\zeta)$ exists at every point of $|\zeta| = 1$, but rather that the entire expression has an analytic continuation to $|\zeta| = 1$.

We could also have obtained the identity of Theorem 10.10 for $|\zeta| < 1$ using the Goluzin interior variation [G7] . For this interior variation of univalent functions in the unit disk, see the elementary derivation of Chr. Pommerenke [P7].

Theorem 10.10 was first proved for the coefficient problem in the class S by M. Schiffer [S2,S5]. In [S5] he introduced a particularly elegant variation of the Green's function under an interior variation of the domain. It applies even to multiply connected domains and is very useful in studying potential-theoretic domain functionals. In general, variations of a domain do not preserve its conformal moduli. So great care must be taken to obtain variations for univalent functions in multiply connected domains. However, no difficulty arises for simply connected domains; variations of Green's functions lead directly to variations for univalent functions. Unfortunately, this important variation will not be treated in these notes.

We should also make note of the Hadamard variation, which is a boundary variation for the Green's function of a multiply connected domain (see S. E. Warschawski [W1]). A consequence of it is the Julia variation [J3], which is a boundary variation for univalent functions in the unit disk.

Since the expressions in Theorem 10.10 are analytic in ζ, we may equate Maclaurin coefficients. The constant coefficients give

$$L_f(f') = \overline{L_f(z^2 f')} \ ,$$

which is Theorem 9.1(B). Of course, the remaining coefficients give an infinity of such relations.

We now restate some of our knowledge for the class S.

<u>COROLLARY 10.11</u>. Suppose $f \in S$, $L \in H'(U)$ is not of the form $L(g) = \alpha g(0) + \beta g'(0)$, and $\operatorname{Re} L(f) = \max_S \operatorname{Re} L$. Then either $f(z) = z/(1-\eta z)^2$, $|\eta| = 1$, or $\mathbb{C}-f(U)$ is a single analytic arc satisfying the differential equation $L(f^2/(f-w))(\frac{dw}{w})^2 > 0$. In any case,

$$L\left(\left[\frac{\zeta f'(\zeta)}{f(\zeta)}\right]^2 \frac{f(z)^2}{f(z)-f(\zeta)} - \frac{\zeta z f'(z)}{z-\zeta}\right) = L(f) - \overline{L\left(\frac{\bar{\zeta} z^2 f'(z)}{1-\bar{\zeta} z}\right)}$$

for $|\zeta| \le 1$.

<u>Proof</u>. The corollary is a consequence of Corollary 10.6 and

Theorem 10.10. If $L(f^2/(f-w))$ vanishes at some point of $\mathbb{C}-f(U)$, then $\mathbb{C}-f(U)$ lies on a straight line through the origin, and consequently, f is a Koebe function by the subordination principle (Theorem 1.1).

Let us interpret the differential equations of Corollary 10.11 in terms of the representation for $L \in H'(U)$ from Corollary 4.2.

COROLLARY 10.11′. Suppose $f \in S$ is extremal for the problem $\max_S \operatorname{Re} \sum_{n=0}^{\infty} a_n b_n$, where $\limsup_{n\to\infty} |b_n|^{1/n} < 1$ and $b_n \neq 0$ for some $n \geq 2$. Then either f is a Koebe mapping or $\mathbb{C}-f(U)$ is a single analytic arc satisfying the differential equation

$$\sum_{n=2}^{\infty} \sum_{k=2}^{n} \frac{a_n^{(k)} b_n}{w^{k+1}} (dw)^2 < 0$$

where $[f(z)]^k = \sum_{n=k}^{\infty} a_n^{(k)} z^n$. In any case,

$$[\zeta f'(\zeta)]^2 \sum_{n=2}^{\infty} \sum_{k=2}^{n} \frac{a_n^{(k)} b_n}{[f(\zeta)]^{k+1}} = \sum_{n=2}^{\infty} (n-1) a_n b_n + \sum_{n=2}^{\infty} \sum_{k=1}^{n-1} k\left(a_k b_n \zeta^{k-n} + \overline{a_k b_n} \zeta^{n-k}\right)$$

for $|\zeta| \leq 1$.

EXERCISE. Give appropriate interpretations to the previous equation for $\zeta = 0$ and $|\zeta| = 1$.

EXERCISE. Show that the Koebe function $k(z) = z/(1-\eta z)^2$, $|\eta| = 1$, satisfies the equation of Corollary 10.11 iff the function

$$F(\zeta) = L\left(\frac{\eta z^2 (1+\eta z)}{(1-\eta z \zeta)(1-\eta z)^3}\right)$$

satisfies $F(\zeta) = \overline{F(\bar{\zeta})}$ (i.e., F has real Maclaurin coefficients).

In particular, the Koebe function $z/(1-z)^2$ satisfies the equation of Corollary 10.11 if $b_n = L(z^n) \in \mathbb{R}$ for all $n \geq 2$ (i.e., L has real coefficients).

Even if L has real coefficients these observations do not solve the problem $\max_S \operatorname{Re} L$, since the equation of Corollary 10.11 is only a necessary condition; it is not necessarily sufficient.

EXAMPLE. If f is extremal for the Bieberbach problem $\max_S \operatorname{Re} a_n$ $(n \geq 2)$, then either f is a Koebe function or the analytic arc $\mathbb{C}-f(U)$ satisfies

$$\sum_{k=2}^{n} \frac{a_n^{(k)}}{w^{k+1}} (dw)^2 < 0 .$$

In any case,

$$[zf'(z)]^2 \sum_{k=2}^{n} \frac{a_n^{(k)}}{[f(z)]^{k+1}} = (n-1)a_n + \sum_{k=1}^{n-1} k(a_k z^{k-n} + \overline{a}_k z^{n-k})$$

for $|z| \leq 1$.

Let us consider some specific linear problems. Let D be an arbitrary plane domain, and consider the problem $\max_{\mathcal{S}(D,z_0)} \operatorname{Re}\{e^{i\alpha} f''(z_0)\}$. By Corollary 10.6, each nondegenerate component γ of $\mathbb{C}-f(D)$ is an analytic arc satisfying

$$L(f^2/(f-w)) \left(\frac{dw}{w}\right)^2 = -\frac{2e^{i\alpha}}{w^3} (dw)^2 > 0 .$$

Since $d(e^{\frac{1}{2}i\alpha}/\sqrt{w})$ is imaginary on γ, we may parametrize γ by

$$w(t) = \left(\frac{1}{\sqrt{w_0}} + ie^{-\frac{1}{2}i\alpha} t\right)^{-2}, \quad w_0 \in \gamma .$$

If γ is unbounded, we may choose $w_0 = \infty$; then γ lies on the ray $\{-te^{i\alpha} : t > 0\}$. We have observed the following:

EXAMPLE. Solutions to the problem $\max_{\mathcal{S}(D,z_0)} \operatorname{Re}\{e^{i\alpha} f''(z_0)\}$ are slit mappings. If γ is a nondegenerate component of $\mathbb{C}-f(D)$ and $w_0 \in \gamma$, then γ lies on the limaçon $w(t) = \left(\frac{1}{\sqrt{w_0}} + ie^{-\frac{1}{2}i\alpha} t\right)^{-2}$. If γ is unbounded (at most one unbounded component exists by Corollary

10.6), then it lies on the ray $\{-te^{i\alpha} : t > 0\}$.

In particular, for the class S the slit for the extremal mapping must lie on the indicated ray. By the subordination principle (Theorem 1.1), the extremal function must be $k(z) = z/(1-e^{-i\alpha}z)^2$, and $\max_S \operatorname{Re}\{e^{i\alpha}a_2\} = 2$.

<u>REMARK</u>. Note that $k(z) = z/(1-\eta z)^2$ was the unique extremal function for the problem $\max_S \operatorname{Re}\{\bar{\eta}a_2\}$, $|\eta| = 1$. It follows from Theorem A.3 that each Koebe function is not only a support point of S but also an extreme point of S . Other extreme points for S can be obtained by posing linear extremal problems over S that have unique solutions.

<u>EXERCISE</u>. Consider the problem $\max_{\Sigma(D,p,q,P,Q)} \operatorname{Re}\{e^{i\alpha}f'(p)\}$. Show that nondegenerate components of $\mathbb{C}-f(D)$, for an extremal function f, lie on arcs of the form

$$w(t) = Q + (P-Q)\left[1 + ce^{ite^{-\frac{1}{2}i\alpha}/\sqrt{f'(p)}}\right]^2 \Big/ \left[1 - ce^{ite^{-\frac{1}{2}i\alpha}/\sqrt{f'(p)}}\right]^2 ,$$

and for an unbounded component $c = 1$.

<u>THEOREM 10.12</u> (Löwner [L6]). $\max_S \operatorname{Re} a_3 = 3$, and $k(z) = z/(1 \pm z)^2$ are the only extremal functions.

<u>Proof</u> (Garabedian and Schiffer [G2]). Let f be an extremal function. If f is a Koebe function, then f must be of the indicated form since their third coefficients have largest real part among all Koebe functions. By Corollary 10.11′ (or the example on page 105) the only alternative is that $\mathbb{C}-f(U)$ is a single analytic arc satisfying

$$(2a_2 w+1)\left(\frac{dw}{w^2}\right)^2 < 0 .$$

If $a_2 = 0$, then $[d(1/w)]^2 < 0$ and $1/w(t) = \int_{\infty}^{w(t)} d(1/w) = \int_0^t i\,dt = it$, so that $\mathbb{C}-f(U)$ is an arc of the imaginary axis. By the subordination principle, $f(z) = z/(1 \pm iz)^2$ and $\Re\, a_3 = 0$, which of course is absurd. We may therefore assume $2a_2 = \rho e^{i\alpha}$, $\rho > 0$.

Instead of $\mathbb{C}-f(U)$, it is convenient to consider the bounded analytic arc γ obtained from $\mathbb{C}-f(U)$ by the inversion $\omega = -1/(2a_2 w) = -1/(\rho e^{i\alpha} w)$. Then γ extends from $\omega = 0$ to a finite point and satisfies the differential equation

$$\left(\frac{1-\omega}{\omega}\right)(e^{i\alpha} d\omega)^2 > 0$$

except at $\omega = 0$. We conclude from this equation that $1 \notin \gamma$ and that

$$\sqrt{\frac{1-\omega}{\omega}}\, d\omega$$

has constant argument on $\gamma - \{0\}$.

Consider the Schwarz - Christoffel mapping

$$W = F(\omega) = \int_0^{\omega} \sqrt{\frac{1-\omega}{\omega}}\, d\omega .$$

To understand the total mapping we note that for one branch of the integrand F maps the upper half-plane as indicated:

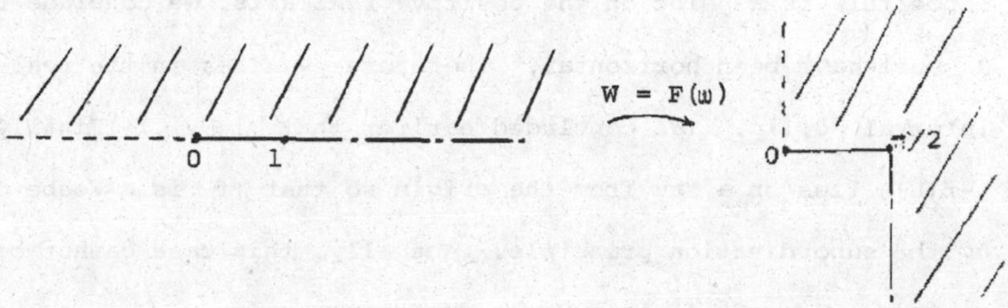

Therefore by reflection:

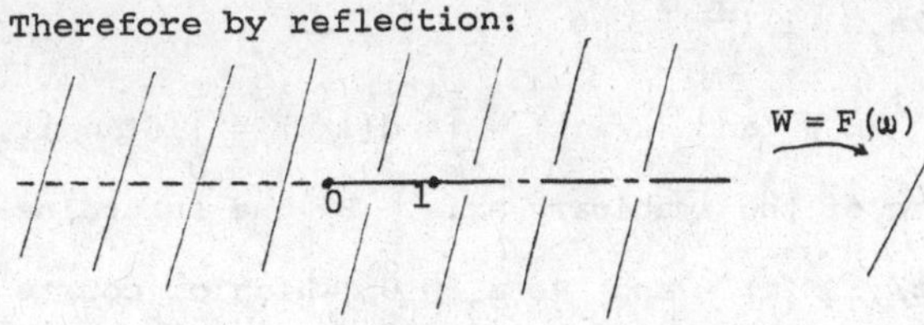

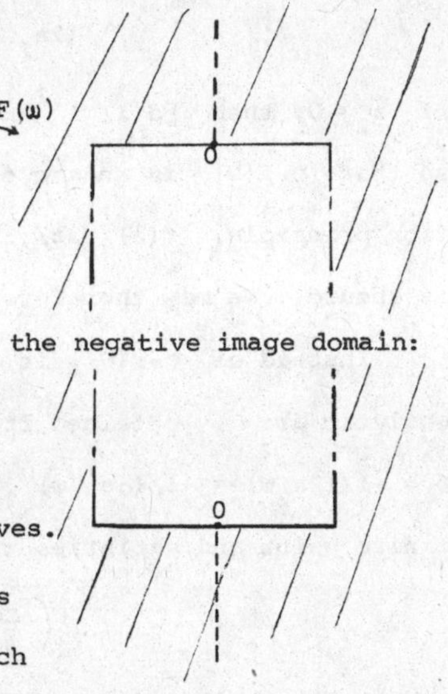

The other branch of the integrand gives the negative image domain: Therefore F maps the ω-plane onto a two-sheeted surface with the indicated boundaries identified with their negatives.

F maps γ onto an arc Γ on this surface, with one endpoint at $W = 0$, such that dW has constant argument on Γ. Therefore Γ is a straight line segment on the surface, with one endpoint at the origin.

If Γ is not horizontal, then its preimage γ lies entirely in the upper half-plane, the lower half-plane, or the negative real axis (except for the endpoint at the origin). We compute the following convex sum of points on γ:

$$\frac{1}{2\pi}\int_0^{2\pi} \frac{-1}{2a_2 f(e^{i\theta})}\, d\theta = \lim_{r\to 1} \frac{1}{2\pi i}\int_{|z|=r} \frac{-1}{2a_2 f(z)}\,\frac{dz}{z} = \operatorname*{Res}_{z=0} \frac{-1}{2a_2 z f(z)} = \frac{1}{2}\,.$$

Since this is a point on the positive real axis, we conclude that Γ must have been horizontal. Therefore γ lies in the real interval $[0,1)$. (We concluded earlier that $1 \notin \gamma$.) Finally then, $C-f(U)$ lies on a ray from the origin, so that f is a Koebe mapping by the subordination principle. Actually, this case cannot occur since $1 \in \gamma$ if f is a Koebe function.

It follows from the rotation $f \to e^{-i\alpha} f(e^{i\alpha} z)$ that $\max_S |a_3| = 3$.

A proof along similar lines that leads to $\max_{S} |a_4| = 4$ has been given by Z. Charzynski and M. Schiffer [C2].

We conclude this chapter with some applications to the class $\Sigma(D)$. Each $g \in \Sigma(D)$ has an expansion $g(z) = z + \sum_{n=0}^{\infty} b_n z^{-n}$ in a neighborhood of ∞. Moreover, $\Sigma'(D)$ is never empty since it contains the identity mapping.

In order to apply Theorem 10.1 we shall need to find $L_g(1/(g-w))$ for $g \in \Sigma'(D)$. One easily verifies that $L_g = L - L(g)\ell_1 - L(1)\ell_2$, where ℓ_1 and ℓ_2 are given in the last example on page 57. It follows that

$$L_g(1/(g-w)) = L(1/(g-w))$$

for $g \in \Sigma'(D)$.

We now consider the linear extremal problem

$$\max_{\Sigma'(D)} \operatorname{Re}\{e^{-2i\alpha} b_1\} .$$

If g is an extremal function, then the differential

$$L_g(1/(g-w))(dw)^2 = L(1/(g-w))(dw)^2 = e^{-2i\alpha}(dw)^2 .$$

By Theorem 10.1 each nondegenerate component γ of $\mathbb{C}-g(D)$ is an analytic arc satisfying the differential equation $e^{-2i\alpha}(dw)^2 > 0$; that is, $e^{-i\alpha}dw$ is real. So γ lies on a straight line with inclination α. We have proved the following:

<u>THEOREM 10.13</u>. Let $\alpha \in [0,\pi)$. Then there exists a $g \in \Sigma'(D)$ that maps D conformally onto a domain whose complement consists entirely of points and line segments with inclination α.

Since every domain is conformally equivalent to a domain bounded by points and slits with inclination α, we call the latter a <u>canonical domain</u>. In the next chapter we shall obtain more canonical domains.

<u>EXERCISE</u>. Verify that $\max_{\Sigma}|b_1| = 1$ and that $z + b_0 + \eta/z$, $|\eta| = 1$, are the only extremal functions.

In the next chapter (Theorem 11.10) we shall show that $\max_{\Sigma}|b_2| = 2/3$. The surprising result $\max_{\Sigma}|b_3| = \frac{1}{2} + e^{-6}$ was obtained by P. R. Garabedian and M. Schiffer [G2]. This disproved an earlier conjecture that $\max_{\Sigma}|b_n| = 2/(n+1)$ for all $n \geq 1$.

Since the extremal function $z(1+z^{-3})^{2/3}$ for the problem $\max_{\Sigma'} \mathrm{Re}\, b_2$ has an image domain whose boundary branches at the origin, it is evident that Lemmas 10.2 and 10.3 do not apply to Σ'. (In fact, $\ell_0(1/(g-w)) \equiv 0$ in this case.) Although branching can occur, we shall show that there are only finitely many analytic arcs. More generally:

<u>THEOREM 10.14</u>. Suppose $g \in \Sigma'(D)$, $L \in H'(D)$ is nonconstant on $\Sigma'(D)$, the support of some representing measure for L does not separate the components of $\mathbb{C}-D$, and $\mathrm{Re}\, L(g) = \max_{\Sigma'(D)} \mathrm{Re}\, L$. If γ is a nondegenerate component of $\mathbb{C}-g(D)$, then γ consists of finitely many analytic arcs satisfying

$$(*) \qquad L(1/(g-w))(dw)^2 > 0 .$$

The only possible points of nonanalyticity or branching for γ are points where $L(1/(g-w))$ vanishes.

<u>Proof</u>. In view of Theorem 10.1 it is sufficient to show that $L_g(1/(g-w)) = L(1/(g-w))$ does not vanish identically on γ. We assume for the purpose of contradiction that it does. Choose R sufficiently large that $\{|z| \geq R\} \subset D$, and define $L_0 \in H'(D)$ by $L_0(h) = \frac{1}{2\pi}\int_0^{2\pi} h(Re^{i\theta})\,d\theta$. For fixed $p \in \gamma$, define $\zeta = \varphi(w) = 1/(w-p)$. Let $\tilde{D} = \varphi \circ g(D) \cup \{0\}$ and define $M \in H'(\tilde{D})$ by $M(\tilde{h}) = (L - L(1)L_0)(\tilde{h} \circ \varphi \circ g)$. Since $L(1/(g-w))$ is analytic in a connected neighborhood of $\mathbb{C}-g(D)$, it vanishes identically on that neighborhood. It follows that

$$M(1/(t-\zeta)) = L(1/(\varphi \circ g - \zeta)) + L(1)/\zeta = -(w-p)^2 L(1/(g-w))$$

vanishes for ζ in a neighborhood of $\bar{\mathbb{C}}-\tilde{D}$. By Corollary 4.4, $M \equiv 0$ on $H(\tilde{D})$. By a change of variables, $L - L(1)L_0$ vanishes for those functions in $H(D)$ that are finite at ∞. If now $h \in \Sigma'(D)$, then $L(h-z) = 0$. Consequently, L is constant on $\Sigma'(D)$, contradicting the hypothesis.

For Σ' we may phrase Theorem 10.14 in terms of support points:

<u>COROLLARY 10.15</u>. Suppose g is a support point of Σ'. Then $\mathbb{C}-g(|z|>1)$ consists of finitely many analytic arcs satisfying (*).

CHAPTER 11. Application to some nonlinear problems

Schiffer's fundamental lemma (Theorem C.4) is also a powerful tool for attacking nonlinear problems:

THEOREM 11.1. Suppose $\mathfrak{F} = \mathfrak{F}(D,\ell_1,\ell_2,P,Q)$ is compact, $f \in \mathfrak{F}$, λ is a real continuous functional on $\mathfrak{F}$ with complex Gâteaux derivative $L \in H'(D)$ at f relative to $\mathfrak{F}$, $\lambda(f) = \max_{\mathfrak{F}} \lambda$, $L_f = L + L(f)\ell_o - L(1)\tilde{\ell}_o$, and γ is a nondegenerate component of $\mathbb{C}-f(D)$. If $L_f(1/(f-w)) \not\equiv 0$ on γ, then γ consists of finitely many analytic arcs each satisfying

$$L_f(1/(f-w))(dw)^2 > 0 .$$

The only possible points of nonanalyticity or branching of γ are zeros of $L_f(1/(f-w))$. Consequently, if $L_f(1/(f-w))$ does not vanish on γ, then γ is a single analytic arc satisfying $L_f(1/(f-w))(dw)^2 > 0$.

Proof. The proof is the same as for Theorem 10.1. Extra $o(\varepsilon)$ terms are not significant.

Let us consider the problem $\max_{\Sigma'(D)} \lambda$ where for fixed $z_o \in D$ and $\alpha \in \mathbb{R}$ the functional λ is defined by $\lambda(g) = \Re\{-e^{-2i\alpha}\log g'(z_o)\}$. This functional is continuous on $\Sigma'(D)$, and its complex Gâteaux derivative L at g is given by $L(h) = -e^{-2i\alpha}h'(z_o)/g'(z_o)$. In this case,

$$L_g(1/(g-w)) = L(1/(g-w)) = e^{-2i\alpha}/[g(z_o)-w]^2 .$$

Suppose g is an extremal function for the problem $\max_{\Sigma'(D)} \lambda$. Then, by Theorem 11.1, each nondegenerate component γ of $\mathbb{C}-g(D)$ is an analytic arc satisfying

$$\left[\frac{e^{-i\alpha}dw}{g(z_o)-w}\right]^2 > 0 .$$

If $w_o \in \gamma$, then γ may be parametrized by

$$w(t) = g(z_o) + [w_o - g(z_o)]e^{e^{i\alpha}t} .$$

For $\alpha = 0$ this arc lies on a ray from $g(z_o)$; for $\alpha = \frac{1}{2}\pi$ it lies on a circle about $g(z_o)$; and for $0 < |\alpha| < \frac{1}{2}\pi$ one obtains the various logarithmic spirals. After a translation we have proved the following:

THEOREM 11.2. There exist $g_j \in \Sigma(D)$, $j = 1,2,3$, such that for fixed $z_o \in D$, $g_j(z_o) = 0$, and $g_1(D)$ is bounded by points and radial slits toward the origin, $g_2(D)$ is bounded by points and circular slits about the origin, and $g_3(D)$ is bounded by points and similar spiral slits about the origin.

As a consequence, the complements of radial slits, circular slits, and spiral slits (plus points) form canonical domains. In a series of articles P. Koebe [e.g, K3-4] discussed quite a number of canonical domains. For example, each finitely connected domain is conformally equivalent to a domain bounded only by points and circles. For a proof by variational methods, see M. Schiffer [S7-9]. This circle normalization is known only for finitely connected domains and certain infinitely connected domains (see K. Strebel [S19]).

PROBLEM (Koebe). Is every plane domain conformally equivalent to a domain bounded only be circles and points? R. J. Sibner [S16] has shown that a domain is conformally equivalent to such a domain iff it is quasiconformally equivalent.

EXERCISE. By considering the problems $\max_{\Sigma'(D)} \min |g(z_1)-g(z_2)|$ for fixed $z_1, z_2 \in D$, prove that there exist $g_j \in \Sigma(D)$, $j = 4,5$, such that $g_4(D)$ is bounded by points and arcs of confocal ellipses centered at the origin and $g_5(D)$ is bounded by points and arcs on confocal hyperbolae centered at the origin.

DEFINITION. Suppose $L \in H'(D)$, φ is analytic in $D \times D$, and $\psi(z,\zeta) = \varphi(z,\bar{\zeta})$. We define

$$L^2(\varphi) = L(L(\varphi)) \quad \text{and} \quad |L|^2(\psi) = L(\overline{L(\psi)})$$

where we compose L successively with the function of the first remaining variable.

For L^2 the order of composition is not important by Fubini's theorem. For $|L|^2$ we note that $|e^{i\alpha}L|^2 = |L|^2$ and that $|L|^2(\psi)$ is real if $\overline{\psi(z,\zeta)} = \psi(\bar{z},\bar{\zeta})$, i.e., if ψ is hermitian.

THEOREM 11.3. Suppose $g \in \Sigma$ and $L \in H'(|z| > 1)$. Then

$$\left|L^2\left(\log \frac{g(z)-g(\zeta)}{z-\zeta}\right)\right| \le |L|^2\left(\log \frac{1}{1-1/(z\bar{\zeta})}\right).$$

Proof. The functional $\lambda(g) = \operatorname{Re} L^2\left(\log \frac{g(z)-g(\zeta)}{z-\zeta}\right)$ is defined and continuous on Σ', and we may consider the problem $\max_{\Sigma'} \lambda$. Let g be an extremal function. Then the complex Gâteaux derivative of λ at g relative to Σ' is

$$L(h;g) = L^2\left(\frac{h(z)-h(\zeta)}{g(z)-g(\zeta)}\right).$$

We compute also

$$L_g(1/(g-w);g) = L(1/(g-w);g) = L^2(-1/([g(z)-w][g(\zeta)-w])) = -[L(1/(g-w))]^2 .$$

If L is the zero functional, the theorem is trivial. Assume therefore that $L \not\equiv 0$. Then by Lemma 4.5, $L(1/(g-w)) \not\equiv 0$ on $C-g(|z|>1)$. Consequently, it follows from Theorem 11.1 that $C-g(|z|>1)$ consists of finitely many analytic arcs satisfying the differential equation

$$[L(1/(g-w))dw]^2 < 0 .$$

Since this is a perfect square, $iL(1/(g-w))dw$ is real.

Let $|\zeta_o| = 1$ and $w_o = g(\zeta_o)$. Then

$$\int_{w_o}^{w} iL(1/(g-w))dw$$

is real if the path is restricted to $C-g(|z|>1)$. We parametrize $C-g(|z|>1)$ by $w = g(\zeta)$, $|\zeta| = 1$. Then

$$G(\zeta) = \int_{w_o}^{g(\zeta)} iL(1/(g-w))dw$$

is real, finite, and continuous for $|\zeta| = 1$. Since a representing measure for L has compact support in $|z| > 1$, G has an analytic extension to $1 < |\zeta| < R$ for some $R > 1$. By the Schwarz reflection principle, G has an analytic continuation across each point of $|\zeta| = 1$. Its derivative

$$G'(\zeta) = ig'(\zeta)L(1/[g(z) - g(\zeta)])$$

therefore also has an analytic continuation across $|\zeta| = 1$, and $i\zeta G'(\zeta)$ is real on $|\zeta| = 1$. By adding $L(\zeta/(z-\zeta)) - \overline{L(1/(1-z\bar{\zeta}))}$, which is also real on $|\zeta| = 1$, to $i\zeta G'(\zeta)$, we find that

$$\Phi(\zeta) = L(\zeta g'(\zeta)/[g(\zeta)-g(z)] + \zeta/(z-\zeta)) - \overline{L(1/(1-z\bar{\zeta}))}$$

is real on $|\zeta|=1$. In fact, $\zeta g'(\zeta)/[g(\zeta)-g(z)] + \zeta/(z-\zeta)$ has a removable singularity at $\zeta = z$. So Φ is defined and analytic for $|\zeta| \geq 1$, real on $|\zeta|=1$, and $\Phi(\infty) = 0$. By the Schwarz reflection principle, Φ extends to a bounded analytic function in $\mathbb{C}$. So by Liouville's Theorem, $\Phi(\zeta) \equiv \Phi(\infty) = 0$.

We divide the identity $\Phi(\zeta) \equiv 0$ by ζ and write it as

$$L\Big(\frac{\partial}{\partial\zeta} \log \frac{g(z)-g(\zeta)}{z-\zeta}\Big) = \overline{L\Big(\frac{\partial}{\partial\bar{\zeta}} \log \frac{1}{1-1/(z\bar{\zeta})}\Big)} .$$

We may interchange L with the derivatives, since L is continuous, and integrate from ∞ to ζ:

$$L\Big(\log \frac{g(z)-g(\zeta)}{z-\zeta}\Big) = \overline{L\Big(\log \frac{1}{1-1/(z\bar{\zeta})}\Big)} .$$

By applying L to both sides, we have

$$\lambda(g) = |L|^2\Big(\log \frac{1}{1-1/(z\bar{\zeta})}\Big) ,$$

where the right side is real since $-\log[1-1/(z\bar{\zeta})]$ is hermitian. Since g was an extremal function,

$$\operatorname{Re} L^2\Big(\log \frac{g(z)-g(\zeta)}{z-\zeta}\Big) \leq |L|^2\Big(\log \frac{1}{1-1/(z\bar{\zeta})}\Big) \quad \text{for all } g \in \Sigma' .$$

This inequality is not affected by adding a constant to g; so it holds for all $g \in \Sigma$. The theorem now follows by replacing L by $e^{i\alpha}L$ for all $\alpha \in \mathbb{R}$.

<u>COROLLARY 11.4</u>. If $f \in S$ and $L \in H'(U)$, then

$$\Big|L^2\Big(\log\Big[\frac{f(z)-f(\zeta)}{z-\zeta}\,\frac{z\zeta}{f(z)f(\zeta)}\Big]\Big)\Big| \leq |L|^2\Big(\log \frac{1}{1-z\bar{\zeta}}\Big) .$$

If $f \in H_u(U)$, $L \in H'(U)$, and $L(1) = 0$, then

$$\Big|L^2\Big(\log \frac{f(z)-f(\zeta)}{z-\zeta}\Big)\Big| \leq |L|^2\Big(\log \frac{1}{1-z\bar{\zeta}}\Big) .$$

Proof. The first inequality follows by applying Theorem 11.3 to $g(z) = 1/f(1/z)$. If, in addition, $L(1) = 0$, then $L^2\left(\log \frac{z\zeta}{f(z)f(\zeta)}\right) = 0$, and the second inequality follows for $f \in S$. Replacing f by $Af+B$ does not affect the second inequality; so it is valid for $f \in H_u(U)$.

We have already tacitly defined the difference quotient $\frac{h(z)-h(\zeta)}{z-\zeta}$ as $h'(\zeta)$ when $z=\zeta$.

COROLLARY 11.5 (Goluzin distortion theorem [G7]). Suppose $g \in \Sigma$ and $f \in S$. Then

$$\left|\sum_{m,n=1}^{N} \lambda_m\lambda_n \log \frac{g(z_m)-g(z_n)}{z_m - z_n}\right| \le \sum_{m,n=1}^{N} \lambda_m\overline{\lambda_n} \log \frac{1}{1-1/(z_m\overline{z_n})}$$

for all $z_n, \lambda_n \in C$ with $|z_n| > 1$ $(n=1, \ldots, N)$, and

$$\left|\sum_{m,n=1}^{N} \lambda_m\lambda_n \log \frac{f(z_m)-f(z_n)}{z_m - z_n} \frac{z_m z_n}{f(z_m)f(z_n)}\right| \le \sum_{m,n=1}^{N} \lambda_m\overline{\lambda_n} \log \frac{1}{1-z_m\overline{z_n}}$$

for all $z_n \in U$ and all $\lambda_n \in C$ $(n=1, \ldots, N)$. In particular,

$$|\log g'(z)| \le \log \frac{1}{1-|z|^{-2}}$$

and

$$\left|\log \frac{z^2 f'(z)}{f(z)^2}\right| \le \log \frac{1}{1-|z|^2}.$$

Proof. Apply Theorem 11.3 and the first part of Corollary 11.4 to the linear functionals $L(h) = \sum_{m=1}^{N} \lambda_m h(z_m)$ and $L(h) = h(z)$.

DEFINITION. The Schwarzian derivative of h is

$$\{h;z\} = 6\left[\frac{\partial^2}{\partial z\partial\zeta} \log \frac{h(z)-h(\zeta)}{z-\zeta}\right]_{\zeta=z} = \left(\frac{h''}{h'}\right)' - \tfrac{1}{2}\left(\frac{h''}{h'}\right)^2$$

It is evident that $\{Ah+B;h\} = \{h;z\}$ and $\{1/h;z\} = \{h;z\}$. Consequently h and $\tau\circ h$ have the same Schwarzian derivative for every Möbius transformation τ. Actually, this property holds more generally. With a little care so that it is defined for $\tau\circ h$, the operation $L^2\left(\log \frac{h(z)-h(\zeta)}{z-\zeta}\right)$ has this property if we assume that h is univalent and $L(1) = 0$. The Schwarzian derivative is obtained by choosing the special functional $L(\varphi) = \sqrt{6}\varphi'(z)$, for fixed z.

COROLLARY 11.6 (Kraus [K6]). If $g \in H_u(|z|>1)$ and $f \in H_u(U)$, then

$$|\{g;z\}| \le 6/(|z|^2-1)^2 \quad \text{and} \quad |\{f;z\}| \le 6/(1-|z|^2)^2 .$$

Proof. Apply Theorem 11.3 and the second part of Corollary 11.4 with the functional $L(\varphi) = \sqrt{6}\,\varphi'(z)$.

COROLLARY 11.7 (Grunsky inequalities [G11]). Let $f \in H_u(U)$ have Grunsky matrix $[c_{mn}]$ generated by

$$\log \frac{f(z)-f(\zeta)}{z-\zeta} = \sum_{m,n=0}^{\infty} c_{mn} z^m \zeta^n$$

and $g \in \Sigma$ have Grunsky matrix $[\gamma_{mn}]$ generated by

$$\log \frac{g(z)-g(\zeta)}{z-\zeta} = \sum_{m,n=1}^{\infty} \gamma_{mn} z^{-m} \zeta^{-n} .$$

Then

$$\left|\sum_{m,n=1}^{\infty} c_{mn}\lambda_m\lambda_n\right| \le \sum_{n=1}^{\infty} \frac{|\lambda_n|^2}{n}$$

and

$$\left|\sum_{m,n=1}^{\infty} \gamma_{mn}\lambda_m\lambda_n\right| \le \sum_{n=1}^{\infty} \frac{|\lambda_n|^2}{n}$$

for all complex sequences $\{\lambda_n\}$ with $\limsup_{n\to\infty} |\lambda_n|^{1/n} < 1$.

<u>Proof</u>. By Theorem 4.2, an $L \in H'(U)$ exists such that $L(z^n) = \lambda_n$, $n \geq 1$, and $L(1) = 0$. For this L, apply the second inequality of Corollary 11.4 to obtain the first inequality above. For the second one, construct an $L \in H'(|z|>1)$ such that $L(z^{-n}) = \lambda_n$, $n \geq 1$, and apply Theorem 11.3.

<u>REMARKS</u>. We deduced the Grunsky inequalities from the inequalities in Corollary 11.4 and Theorem 11.3, respectively. They are, in fact, <u>equivalent</u> to these inequalities. This is easily seen by reversing the construction in the previous proof.

H. Grunsky [G11] obtained inequalities of the above type even for finitely connected domains.

<u>COROLLARY 11.8</u>. The Grunsky inequalities of Corollary 11.7 are not only necessary, but also sufficient, for $f \in H(U)$, $f'(0) \neq 0$, and $g \in H(|z|>1)$, $g'(\infty) = 1$, to be univalent.

<u>Proof</u>. The necessity is Corollary 11.7. For the sufficiency we use the equivalence of the Grunsky inequalities with the inequalities of Corollary 11.4 and Theorem 11.3. The conditions $f(z_1) = f(z_2)$ or $g(z_1) = g(z_2)$ for $z_1 \neq z_2$ will violate the latter inequalities for obvious choices of L that render the left side infinite.

<u>COROLLARY 11.9</u>. Let $f \in H_u(U)$ have Grunsky matrix $[c_{mn}]$ and $g \in \Sigma$ have Grunsky matrix $[\gamma_{mn}]$. Then

$$\sum_{m=1}^{N} m \left| \sum_{n=1}^{N} c_{mn} \lambda_n \right|^2 \leq \sum_{n=1}^{N} \frac{|\lambda_n|^2}{n} \quad \text{and} \quad \sum_{m=1}^{N} m \left| \sum_{n=1}^{N} \gamma_{mn} \lambda_n \right|^2 \leq \sum_{n=1}^{N} \frac{|\lambda_n|^2}{n}$$

for all $\lambda_1, \ldots, \lambda_N \in \mathbb{C}$.

<u>Proof</u>. The finite matrix $C = [\sqrt{mn}\, c_{mn}]_{m,n=1,\ldots,N}$ is (complex) symmetric. By a lemma of I. Schur [S15] there exists a unitary matrix $\mathfrak{U}$ such that $\mathfrak{U}^t C \mathfrak{U}$ is a nonnegative (all $d_n \geq 0$) diagonal matrix $D = \mathrm{diag}(d_1, \ldots, d_N)$. If we write $x^t = (\lambda_1/\sqrt{1}, \ldots, \lambda_N/\sqrt{N})$, then the Grunsky inequalities of Corollary 11.7 become $|x^t Cx| \leq \|x\|^2$. With $x = \mathfrak{U}y$, it follows that

$$|y^t Dy| = |x^t Cx| \leq \|x\|^2 = \|y\|^2$$

for any choice of y. Therefore $0 \leq d_n \leq 1$, $n = 1, \ldots, N$. Now

$$\sum_{m=1}^{N} m \Big|\sum_{n=1}^{N} c_{mn}\lambda_n\Big|^2 = \|Cx\|^2 = \bar{x}^t \bar{C}^t Cx = \bar{x}^t \mathfrak{U}\bar{D}D\bar{\mathfrak{U}}^t x \leq \|x\|^2 = \sum_{n=1}^{N} \frac{|\lambda_n|^2}{n} .$$

The proof for $[\gamma_{mn}]$ is identical.

<u>EXAMPLE</u>. Let $g(z) = z + \sum_{n=0}^{\infty} b_n z^{-n} \in \Sigma$. Then the Grunsky coefficients

$$\sum_{m=1}^{\infty} \gamma_{m1} z^{-m} = \lim_{\zeta\to\infty} \zeta \log \frac{g(z)-g(\zeta)}{z-\zeta} = -\sum_{n=1}^{\infty} b_n z^{-n} ,$$

i.e., $\gamma_{m1} = -b_m$. Therefore Corollary 11.9 with $\lambda_1 = 1$, $\lambda_2 = \ldots = \lambda_N = 0$, becomes $\sum_{m=1}^{N} m|b_m|^2 \leq 1$. By letting $N \to \infty$, we obtain the area theorem (Theorem B.1)

$$\sum_{m=1}^{\infty} m|b_m|^2 \leq 1 .$$

In the proofs of the next two theorems it is apparent that the Grunsky inequalities are a powerful means for obtaining coefficient results.

<u>THEOREM 11.10</u> (Schiffer [S4]). If $g \in \Sigma$ and $g(z) = z + \sum_{n=0}^{\infty} b_n z^{-n}$, then

$$|b_2| \leq 2/3 .$$

<u>Proof</u>. It is no loss of generality to assume that $g \in \Sigma'$ and its $b_2 > 0$. Then

$$\frac{1}{2\pi}\int_0^{2\pi} g(\rho e^{i\theta})\,d\theta = b_o = 0 \qquad \text{for all } \rho > 1 .$$

Consequently, the set $\mathbb{C}-g(|z|>1)$ contains the origin in its closed convex hull. Therefore the set $\mathbb{C}-g(|z|>1)$ contains a point c in one of the three sectors defined by $\mathrm{Re}\, z^3 \le 0$. Now apply the Grunsky inequalities of Corollary 11.7 with $\lambda_3 = 1$, $\lambda_n = 0$ for $n \ne 3$, to $\sqrt{g(z^2)-c} \in \Sigma$. Then

$$\mathrm{Re}\{\tfrac{1}{2}b_2 - \tfrac{1}{48}c^3\} = \mathrm{Re}\{-\gamma_{33}\} \le 1/3 .$$

Since $\mathrm{Re}\{c^3\} \le 0$, one has $b_2 \le 2/3$.

<u>EXERCISE</u>. Verify that the only function in Σ' with $b_2 = 2/3$ is $z(1+z^{-3})^{2/3}$.

<u>THEOREM 11.11</u> (Garabedian and Schiffer [G3]). $\max\limits_S |a_4| = 4$, and $k(z) = z/(1-\eta z)^2$, $|\eta| = 1$, are the only extremal functions.

<u>Proof</u> (Charzynski and Schiffer [C3]). Let $f(z) = z + \sum_{n=2}^{\infty} a_n z^n \in S$. Then

$$z/\sqrt{z^2 f(1/z^2)} = z - \tfrac{1}{2}a_2\frac{1}{z} - \tfrac{1}{2}(a_3 - \tfrac{3}{4}a_2^2)\frac{1}{z^3} - \dots \in \Sigma .$$

So by the area theorem (Theorem B.1 or the previous example), we have

$$|\tfrac{1}{2}a_2|^2 + 3|\tfrac{1}{2}(a_3 - \tfrac{3}{4}a_2^2)|^2 \le 1$$

and $|a_3 - \frac{3}{4}a_2^2| \le \frac{1}{\sqrt{3}}\sqrt{4-|a_2|^2}$. We employ the Grunsky inequalities of Corollary 11.7 with $\lambda_1 = \lambda$, $\lambda_3 = 1$, and all other $\lambda_n = 0$:

$$\mathrm{Re}\{\lambda^2 c_{11} + 2\lambda c_{13} + c_{33}\} \le |\lambda|^2 + \tfrac{1}{3} .$$

However, we apply them to the odd function $z\sqrt{f(z^2)/z^2} \in S$. Then

$c_{11} = \frac{1}{2}a_2$, $c_{13} = \frac{1}{2}(a_3 - \frac{3}{4}a_2^2)$, $c_{33} = \frac{1}{2}(a_4 - 2a_2a_3 + \frac{13}{12}a_2^3)$,

so that

$$\begin{aligned} \operatorname{Re} a_4 &\le 2|\lambda|^2 + \frac{2}{3} + \operatorname{Re}\{4(a_2-\lambda)c_{13} + \frac{5}{12}a_2^3 - \lambda^2 a_2\} . \\ &\le 2|\lambda|^2 + \frac{2}{3} + \frac{2}{\sqrt{3}}|a_2-\lambda|\sqrt{4-|a_2|^2} + \operatorname{Re}\{\frac{5}{12}a_2^3 - \lambda^2 a_2\} . \end{aligned}$$

By Theorem B.2 (or the second example on page 105) we may represent $a_2 = 2xe^{i\varphi}$, $0 \le x \le 1$. Now choose $\lambda = 2xe^{-i\varphi/2}\cos\frac{3}{2}\varphi$, and set $y = |\sin\frac{3}{2}\varphi|$. Then

$$\operatorname{Re} a_4 \le -\frac{4}{3}x^2(6-x)y^2 + \frac{8}{\sqrt{3}}x\sqrt{1-x^2}\,y + \frac{2}{3} + 8x^2 - \frac{14}{3}x^3 .$$

The maximum of $-Ay^2 + By$, $A > 0$, is $B^2/(4A)$. Therefore

$$\operatorname{Re} a_4 \le \frac{4(1-x^2)}{6-x} + \frac{2}{3} + 8x^2 - \frac{14}{3}x^3 .$$

The latter expression is at most 4 iff

$$12(1-x^2) + (6-x)(-10 + 24x^2 - 14x^3) \le 0$$

or

$$(48 + 86x - 8x^2)(1-x)^2 + 6x^3(1-x) \ge 0 .$$

This final inequality is obviously satisfied for $0 \le x \le 1$. Equality occurs iff $x = 1$, i.e., when f is a Koebe function.

The substitution $f \to e^{-i\alpha}f(e^{i\alpha}z)$ implies $|a_4| \le 4$; equality is possible only for Koebe functions $k(z) = z/(1-\eta z)^2$, $|\eta| = 1$.

We now turn to the problem of removing the logarithm in Theorem 11.3 and Corollary 11.4. We shall follow a development by S. Friedland [F2]. Similar considerations appear in the work of I. Schur [S13] and Chr. Pommerenke [P8] .

In the following, unless otherwise mentioned, all matrices are assumed to be n by n with complex entries. As on page 120, bars denote complex conjugates, t's denote transposes, and we shall use x and y for n by 1 matrices of complex numbers.

LEMMA 11.12. If A is a positive definite hermitian matrix and B is a symmetric matrix, then there exist a nonsingular matrix R and a nonnegative (all $d_k \geq 0$) diagonal matrix $D = \operatorname{diag}(d_1, \ldots, d_n)$ such that

$$A = \bar{R}^t R \quad \text{and} \quad B = R^t D R .$$

Moreover, $$\max_k d_k = \max_x \frac{|x^t B x|}{\bar{x}^t A x} .$$

Proof. Since A is positive definite, $A = \bar{C}^t C$ for some nonsingular matrix C. The matrix $(C^{-1})^t B C^{-1}$ is symmetric. By a lemma of I. Schur [S15] (used also in the proof of Corollary 11.9), there exists a unitary matrix U such that $U^t (C^{-1})^t B C^{-1} U$ is a nonnegative diagonal matrix $D = \operatorname{diag}(d_1, \ldots, d_n)$. The matrix $R = U^{-1} C$ has the properties $\bar{R}^t R = A$ and $R^t D R = B$. Also,

$$\max_k d_k = \max_y \frac{|y^t D y|}{\bar{y}^t y} = \max_x \frac{|x^t B x|}{\bar{x}^t A x}$$

by substituting $y = Rx$.

DEFINITION. The Kronecker product (direct product, tensor product) of n by n matrices $A = [a_{ij}]$ and $B = [b_{ij}]$ is the n^2 by n^2 matrix

$$A \otimes B = \begin{pmatrix} a_{11}B & \ldots & a_{1n}B \\ \vdots & & \vdots \\ a_{n1}B & \ldots & a_{nn}B \end{pmatrix} .$$

We leave as an exercise several elementary properties of this product:

EXERCISE. (a) $(A \otimes B)(C \otimes D) = (AC) \otimes (BD)$

(b) $(A \otimes B)^t = A^t \otimes B^t$

(c) $\overline{A \otimes B} = \bar{A} \otimes \bar{B}$

DEFINITION. We define $|B| \le A$ to mean that $|x^t Bx| \le \bar{x}^t Ax$ for all x.

If A is a hermitian matrix and B is a symmetric matrix, then $|B| \le A$ means that the hermitian form in A dominates the symmetric form in B. We shall see that this property persists under both Kronecker and Schur multiplication.

LEMMA 11.13. Suppose A_j is hermitian, B_j is symmetric, and $|B_j| \le A_j$, $j = 1,2$. Then $|B_1 \otimes B_2| \le A_1 \otimes A_2$.

Proof. Let $A_j(\varepsilon) = A_j + \varepsilon I$, $\varepsilon > 0$. Then $|B_j| \le A_j(\varepsilon)$ and $A_j(\varepsilon)$ is positive definite. By Lemma 11.12, we have

$$A_j(\varepsilon) = \bar{R}_j^t R_j \quad \text{and} \quad B_j = R_j^t D_j R_j$$

where $D_j = \text{diag}(d_1^{(j)}, \ldots, d_n^{(j)})$ and all $0 \le d_k^{(j)} \le 1$ since $|B_j| \le A_j(\varepsilon)$. Let $R = R_1 \otimes R_2$ and $D = D_1 \otimes D_2$. Then, using properties of the exercise, we have

$$\bar{R}^t R = (\overline{R_1 \otimes R_2})^t (R_1 \otimes R_2) = (\bar{R}_1^t R_1) \otimes (\bar{R}_2^t R_2) = A_1(\varepsilon) \otimes A_2(\varepsilon) ,$$

and, similarly,

$$R^t DR = B_1 \otimes B_2 .$$

D is a diagonal matrix whose diagonal entries $d_k^{(1)} d_\ell^{(2)}$ are non-negative and at most one. Thus $|B_1 \otimes B_2| \le A_1(\varepsilon) \otimes A_2(\varepsilon)$. Let $\varepsilon \to 0$.

<u>DEFINITION</u>. The <u>Schur product</u> (Hadamard product) of n by n matrices $A = [a_{ij}]$ and $B = [b_{ij}]$ is the n by n matrix $A * B = [a_{ij} b_{ij}]$.

$A * B$ is a principal submatrix of $A \otimes B$ obtained by taking the 1st, $n+2^{nd}$, $2n+3^{rd}$, $3n+4^{th}$, ..., $(n-1)n+n^{th} = n^2$ row and column. An immediate corollary of Lemma 11.13 is then:

<u>LEMMA 11.14</u>. Suppose A_j is hermitian, B_j is symmetric, and $|B_j| \le A_j$, $j = 1,2$. Then $|B_1 * B_2| \le A_1 * A_2$.

<u>DEFINITIONS</u>. Associated with a power series $\varphi(z) = \sum_{n=0}^{\infty} c_n z^n$, we define $\varphi^+(z) = \sum_{n=0}^{\infty} |c_n| z^n$. In addition, if a function φ is defined at each element of a matrix $A = [a_{ij}]$, we define the matrix $\varphi * A = [\varphi(a_{ij})]$.

<u>THEOREM 11.15</u> (Friedland [F2], Pommerenke [P8]). Suppose A is hermitian, B is symmetric, $|B| \le A$, and φ is entire. Then $|\varphi * B| \le \varphi^+ * A$.

<u>Proof</u>. $|\underbrace{B * \ldots * B}_{n}| \le \underbrace{A * \ldots * A}_{n}$ by Lemma 11.14. If $\varphi(z) = \sum_{n=0}^{\infty} c_n z^n$, then

$$|\varphi * B| \le \Big|\sum_{n=0}^{\infty} c_n (B * \ldots * B)\Big| \le \sum_{n=0}^{\infty} |c_n| |B * \ldots * B| \le \sum_{n=0}^{\infty} |c_n| (A * \ldots * A) = \varphi^+ * A .$$

<u>THEOREM 11.16</u> ([F2]). Suppose $A(z,\zeta)$ and $B(z,\zeta)$ are analytic in $D \times D$, $A(z,\bar{\zeta}) = \overline{A(\zeta,\bar{z})}$, and $B(z,\zeta) = B(\zeta,z)$. If $|L^2(B(z,\zeta))| \le |L|^2(A(z,\bar{\zeta}))$ for all $L \in H'(D)$, then $|L^2(\varphi \circ B(z,\zeta))| \le |L|^2(\varphi^+ \circ A(z,\bar{\zeta}))$ for all $L \in H'(D)$ and all entire functions φ.

<u>Proof</u>. If $L \in H'(D)$, then by Corollary 4.3 we may represent $L(h) = \int_K h\,d\mu$ for some measure μ with compact support $K \subset D$. Let

μ_n be measures with finite support $K_n = \{z_1, \ldots, z_n\} \subset K$ such that

$$L_n(h) = \int_{K_n} h\,d\mu_n \to \int_K h\,d\mu = L(h) \quad \text{as} \quad n \to \infty ,$$

for each $h \in H(D)$. The matrix $B = [B(z_i, z_j)]$ is symmetric, $A = [A(z_i, \bar{z}_j)]$ is hermitian, and the hypothesis implies $|B| \le A$. By Theorem 11.15, $|\varphi * B| \le \varphi^+ * A$. Consequently, $|L_n^2(\varphi \circ B(z,\zeta))| \le |L_n|^2(\varphi^+ \circ A(z,\bar{\zeta}))$ for each n. Let $n \to \infty$.

In Theorems 11.15 and 11.16 it was not necessary to assume that φ is entire. It would be sufficient to assume that φ is analytic in some open disk centered at the origin that contains the values (elements) of A and B.

We now apply Theorem 11.16 with $\varphi(z) = e^{pz}$ to the inequalities of Theorem 11.3 and Corollary 11.4:

<u>COROLLARY 11.17</u> (FitzGerald [F1]). Suppose $g \in \Sigma$, $f \in S$, and $p \in \mathbb{C}$. Then

$$|L^2([\frac{g(z)-g(\zeta)}{z-\zeta}]^p)| \le |L|^2([1 - \frac{1}{z\bar{\zeta}}]^{-|p|})$$

for all $L \in H'(|z|>1)$, and

$$|L^2([\frac{f(z)-f(\zeta)}{z-\zeta} \frac{z\zeta}{f(z)f(\zeta)}]^p)| \le |L|^2([1 - z\bar{\zeta}]^{-|p|})$$

for all $L \in H'(U)$.

<u>EXERCISE</u>. Obtain inequalities similar to Corollaries 11.5 and 11.7 from Corollary 11.17.

<u>EXERCISE</u>. Define $\|B\| \le A$ to mean that $|\bar{x}^t Bx| \le \bar{x}^t Ax$ for all x. If A and B are hermitian matrices, φ is entire, and

$\|B\| \le A$, prove that $\|\varphi * B\| \le \varphi^+ * A$ (Schur [S13] and Friedland [F2]) .

EXERCISE. Suppose $A(z,\zeta)$ and $B(z,\zeta)$ are analytic in a domain $D \times D$, $A(z,\bar\zeta) = \overline{A(\zeta,\bar z)}$, $B(z,\bar\zeta) = \overline{B(\zeta,\bar z)}$, and φ is entire. If $\left| |L|^2(B(z,\bar\zeta)) \right| \le |L|^2(A(z,\bar\zeta))$ for all $L \in H'(D)$, show that $\left| |L|^2(\varphi \circ B(z,\bar\zeta)) \right| \le |L|^2(\varphi^+ \circ A(z,\bar\zeta))$ for all $L \in H'(D)$.

REMARKS. The inequalities of Corollary 11.17 were originally obtained by C. FitzGerald [F1], using a representation for (a dense set of) univalent functions through the Löwner differential equation [L6]. However, Theorem 11.16 shows that they are a purely algebraic consequence of the inequalities of Theorem 11.3 and Corollary 11.4, which would therefore seem to be the more fundamental entities.

FitzGerald [F1] also obtained and used a related set of inequalities in an ingenious fashion to prove that $|a_n| < \sqrt{7/6}\, n < (1.081)n$ for all n , whenever $f(z) = z + \sum_{n=2}^{\infty} a_n z^n \in S$.

PROBLEM. Use Theorem 11.16, the previous exercise, and ingenuity to uncover further the secrets in the inequalities of Theorem 11.3 and Corollary 11.4.

CHAPTER 12. Some properties of quasiconformal mappings

Let $f = u + iv$ be an orientation preserving homeomorphism of a domain $D \subset \mathbb{C}$. Assume that f is differentiable at $z_o \in D$. Then its Jacobian determinant

$$J_f = \det \begin{vmatrix} u_x & v_x \\ u_y & v_y \end{vmatrix} = |f_z|^2 - |f_{\bar{z}}|^2$$

is positive at z_o. Here we have used the formal derivatives

$$f_z = \tfrac{1}{2}(f_x - if_y) \quad \text{and} \quad f_{\bar{z}} = \tfrac{1}{2}(f_x + if_y),$$

where f_x and f_y are the standard partial derivatives. Near z_o

$$f(z) = f(z_o) + f_z(z_o)(z-z_o) + f_{\bar{z}}(z_o)\overline{(z-z_o)} + o(|z-z_o|)$$

and the linear part or differential

$$df(z_o)(\zeta) = f_z(z_o)\zeta + f_{\bar{z}}(z_o)\bar{\zeta}$$

is an affine transformation that maps circles $|\zeta| = r$ onto ellipses

$$df(z_o)(re^{i\theta}) = f_z(z_o)re^{i\theta} + f_{\bar{z}}(z_o)re^{-i\theta}, \quad 0 \le \theta \le 2\pi .$$

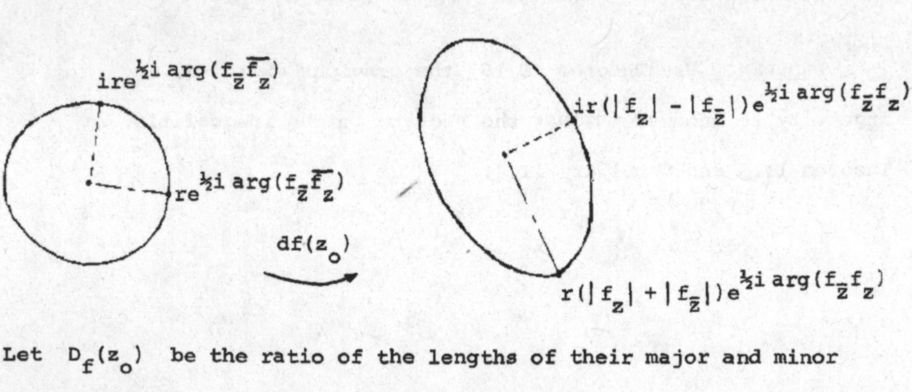

Let $D_f(z_o)$ be the ratio of the lengths of their major and minor axes. Then

$$D_f = \frac{|f_z| + |f_{\bar{z}}|}{|f_z| - |f_{\bar{z}}|} = \frac{(|f_z| + |f_{\bar{z}}|)^2}{|f_z|^2 - |f_{\bar{z}}|^2} = \frac{\|df\|^2}{J_f}$$

We may use $D_f(z_o)$ as a measure of the local distortion of the mapping f at z_o. It is called the dilatation quotient of f at z_o.

We shall need a mild regularity condition:

DEFINITION. A function f is ACL (absolutely continuous on lines) in a domain D if, relative to every closed rectangle $R \subset D$ with sides parallel to the coordinate axes, f is absolutely continuous on a.e. horizontal and vertical line in R.

If f is ACL, then its partial derivatives exist a.e., so that f_z and $f_{\bar{z}}$ are defined a.e.

ANALYTIC DEFINITION OF QUASICONFORMALITY (qcty). Let f be an orientation preserving homeomorphism of a domain $D \subset \mathbb{C}$ into $\bar{\mathbb{C}}$. Assume

(i) f is ACL in D, and

(ii) $D_f \leq K < \infty$ a.e. in D.

Then f is K-quasiconformal (K-q.c.) in D.

It is a remarkable result of F. W. Gehring and O. Lehto [G6] that open mappings, hence homeomorphisms, with partial derivatives a.e. are differentiable a.e. Consequently, q.c. mappings are differentiable a.e.

We may interpret the dilatation condition (ii) to say that infinitessimal circles at a.e. point are mapped onto infinitessimal ellipses whose major axis to minor axis ratio is uniformly bounded by K. In this sense K is a uniform bound for the local distortion.

We are now in the unenviable position of having two definitions for qcty (see also p. 53). We shall remedy this situation. We may restate condition (ii) as $\|df\|^2 \le KJ_f$. So if f is of class C^1 and K-q.c. according to the analytic definition, then Theorem 6.2 implies that $K^{-1}M(\Gamma) \le M(f(\Gamma)) \le KM(\Gamma)$ for all curve families Γ in D. In fact, the ACL property (i) is sufficient smoothness to make the proof of Theorem 6.2 valid. Therefore a K-q.c. mapping according to the analytic definition is also a K-q.c. mapping according to the definition on p. 53 in terms of quasi-invariance of the moduli of curve families. The converse is also true. We shall not pause for the proof, but refer to the standard texts of O. Lehto and K. I. Virtanen [L5] and L. V. Ahlfors [A5]. Therefore the two definitions are completely equivalent, and we shall consider them interchangeably. Actually, there are many equivalent definitions involving the quasi-invariance of the moduli of quadrilaterals (geometric definition), or the moduli of rings, or of angles, etc. (see Gehring [G5]).

The function

$$\mu = f_{\bar{z}}/f_z$$

is called the <u>complex</u> <u>dilatation</u> of f, and

$$D_f \le K \quad \text{iff} \quad |\mu| \le k = (K-1)/(K+1) .$$

So the magnitude of μ is also a measure of the local distortion, and, moreover,

$$\tfrac{1}{2} \arg \mu$$

is the local direction of maximum distortion. Therefore a K-q.c. mapping satisfies the differential equation

$$f_{\bar{z}} = \mu f_z$$

where μ is a measurable function satisfying

$$\|\mu\|_\infty = \operatorname*{ess\,sup}_D |\mu| \leq k \ .$$

Such an equation is called a <u>Beltrami equation</u>.

The function

$$\nu = f_{\bar{z}}/\overline{f_z}$$

is called the <u>second complex dilatation</u> of f, and $|\nu| = |\mu|$ is the same measure of local distortion. However,

$$\tfrac{1}{2} \arg \nu$$

is the local direction of maximum distortion in the image domain. We also see that a K-q.c. mapping satisfies the second differential equation

$$f_{\bar{z}} = \nu \overline{f_z}$$

where ν is a measurable function satisfying

$$\|\nu\|_\infty \leq k = (K-1)/(K+1) \ .$$

It is obvious from the modulus definition of qcty that the inverse of a K-q.c. mapping is also K-q.c. In addition, one sees both from the local geomety and from the identities $(f^{-1})_w = \overline{f_z}/J_f$, $(f^{-1})_{\bar{w}} = -f_{\bar{z}}/J_f$ that

$$\mu_{f^{-1}} = -\nu_f \quad \text{and} \quad \nu_{f^{-1}} = -\mu_f$$

at corresponding points.

We shall see that it is possible to produce mappings for which the local magnitude and direction of maximum distortion are prescribed. That is, it is always possible to find q.c. solutions of the Beltrami equation. It is evident that the theory for the second equation follows from the Beltrami equation for f^{-1}.

PROPOSITION 12.1 (existence and uniqueness theorem). Assume μ is measurable in a domain D with $\|\mu\|_\infty < 1$. Then there exists in D a q.c. mapping f whose derivatives are L^2-integrable on compact subsets and satisfy the Beltrami equation

$$f_{\bar z} = \mu f_z \qquad \text{a.e.}$$

Moreover, every other q.c. mapping of D that satisfies this equation a.e. is of the form $\varphi \circ f$ where φ is conformal.

For proofs of the existence and uniqueness theorem and the facts sketched below we refer the reader again to the standard text of Lehto and Virtanen [L5].

It will be useful to have some representation for q.c. mappings with prescribed complex dilatation. It is sufficient to find one such mapping since all others are then obtained by conformal composition. Moreover, we may as well assume that $D = \mathbb{C}$ since we can extend μ to be zero outside D.

Suppose therefore that μ is measurable in $\mathbb{C}$ and $\|\mu\|_\infty = k<1$. We assume that μ has compact support. Then $\mu = 0$ in a neighborhood of ∞, so the solutions f of the Beltrami equation $f_{\bar z} = \mu f_z$ will be conformal there. By composition with a Möbius transformation we may assume that

$$f(z) = z + o(1) \quad \text{as} \quad z \to \infty .$$

We shall use the Cauchy representation

$$\varphi(z) = \frac{1}{2\pi i} \int_C \frac{\varphi(\zeta)\,d\zeta}{\zeta - z} - \frac{1}{\pi} \iint_{\mathrm{Int}\, C} \frac{\varphi_{\bar\zeta}(\zeta)}{\zeta - z}\, d\xi\, d\eta , \quad z \in \mathrm{Int}\, C ,$$

for smooth functions φ inside and on a rectifiable Jordan curve C. If φ also has compact support, then the first integral goes away if C is sufficiently large, so that

$$\varphi = C\varphi_{\bar{z}} \quad \text{and} \quad \varphi_z = \mathcal{H}\varphi_{\bar{z}}$$

where C and $\mathcal{H}$ are the <u>complex Cauchy and Hilbert transforms</u> defined by

$$C\psi(z) = -\frac{1}{\pi}\iint_{\mathbb{C}} \frac{\psi(\zeta)}{\zeta - z}\, d\xi d\eta \quad \text{and} \quad \mathcal{H}\psi(z) = -\frac{1}{\pi}\iint_{\mathbb{C}} \frac{\psi(\zeta)}{(\zeta - z)^2} d\xi d\eta .$$

At first, we interpret the latter integral as a Cauchy principal value.

Actually, the first formula $\varphi = C\varphi_{\bar{z}}$ holds already if φ is continuous, is ACL, has locally integrable derivatives, and vanishes smoothly at ∞. In particular, for our solution f of the Beltrami equation we have

$$f - z = Cf_{\bar{z}} .$$

Furthermore, the Hilbert transform $\mathcal{H}$ can be extended to $L^p(\mathbb{C})$. Its norm $\|\mathcal{H}\|_p$ is a continuous, strictly increasing function of $p \in [1,\infty)$, and $\|\mathcal{H}\|_2 = 1$. The formula $\varphi_z = \mathcal{H}\varphi_{\bar{z}}$ is still valid if φ is continuous and ACL and its partial derivatives are in $L^2(\mathbb{C})$. In particular, for $\varphi = f - z$ we have

$$f_z - 1 = \mathcal{H}f_{\bar{z}} .$$

So $f_{\bar{z}} = \mu f_z = \mu + \mu\mathcal{H}f_{\bar{z}}$, and

$$(I - \mu\mathcal{H})f_{\bar{z}} = \mu .$$

Since $\|\mu\mathcal{H}\|_2 \le \|\mu\|_\infty \|\mathcal{H}\|_2 = \|\mu\|_\infty < 1$,

$$f_{\bar{z}} = (I - \mu\mathcal{H})^{-1}\mu$$

is a valid representation in $L^2(\mathbb{C})$. In fact, $\|\mu H\|_p \le \|\mu\|_\infty \|H\|_p < 1$, for some p's larger than 2, and for these p's the representation is valid in $L^p(\mathbb{C})$. Finally, by substitution we have the representation

$$f = z + C(I - \mu H)^{-1}\mu .$$

We summarize:

PROPOSITION 12.2 (representation theorem). Let μ be measurable in $\mathbb{C}$ with compact support and $\|\mu\|_\infty < 1$. Then the q.c. mapping f of the plane satisfying

$$f_{\bar{z}} = \mu f_z \quad \text{a.e.}$$

with normalization

$$f(z) = z + o(1) \quad \text{as} \quad z \to \infty$$

has the representation

$$f = z + C(I - \mu H)^{-1}\mu .$$

Actually, the existence part of Proposition 12.1 can be fashioned from this representation.

We mentioned above that $f_{\bar{z}}$ is even L^p integrable for a range of p's larger than 2, depending only on $\|\mu\|_\infty$, i.e., if f is K-q.c., then $f_{\bar{z}} \in L^p(\mathbb{C})$ for $2 \le p < \mathfrak{p}(K)$ where $\|H\|_{\mathfrak{p}(K)} > (K+1)/(K-1)$. It follows from the relation $f_z = 1 + Hf_{\bar{z}}$ that $f_z \in L^p$ over compact sets for the same p's. On the basis of Proposition 12.1, we may compose with some conformal mapping to conclude the following:

PROPERTY 12.3. The derivatives of a K-q.c. mapping f of a domain D belong to L^p over compact subsets of $D - f^{-1}(\{\infty\})$ for $2 \le p < \mathfrak{p}(K)$. It is known that the optimal choice of $\mathfrak{p}(K)$ for each K satisfies

$$\lim_{K\to 1} \mathfrak{p}(K) = \infty \quad \text{and} \quad \lim_{K\to\infty} \mathfrak{p}(K) = 2 .$$

PROPERTY 12.4. It is a consequence of Property 12.3 that q.c. mappings preserve sets of positive measure and sets of measure zero. In particular, if f is a K-q.c. mapping of a domain D , then

$$m(f(E)) = \int_E J_f \, dxdy \le \left(\int_E J_f^{p/2} \, dxdy\right)^{2/p} (m(E))^{\delta}$$

is valid for all compact subsets of $D - f^{-1}(\{\infty\})$ with $\delta = 1 - \frac{2}{p}$, $2 < p < \mathfrak{p}(K)$.

One also has some smoothness properties:

PROPERTY 12.5. Suppose f is a K-q.c. mapping of a domain D , with complex dilatations μ_f and ν_f . Then f is uniformly Hölder continuous, with exponent $1/K$, on compact subsets of $D - f^{-1}(\{\infty\})$. Moreover, if μ_f or ν_f has partial derivatives of order n , $n \ge 0$, that satisfy a Hölder condition with exponent α , then f has partial derivatives or order $n+1$ that satisfy a Hölder condition with exponent α (Agmon, Douglis, and Nirenberg [Al]).

Together with constants, K-q.c. mappings form a closed family, just as for conformal mappings:

PROPERTY 12.6. If f_n is a sequence of K-q.c. mappings of a domain D , converging locally uniformly to f , then f is again a K-q.c. mapping or a constant. Actually more is true: If f_n is a sequence of K-q.c. mappings converging pointwise to f , then either f is K-q.c. and the convergence is locally uniform or f assumes at most two values. For mappings into $\bar{\mathbb{C}}$ we mean convergence in the spherical metric.

Theorem 6.6 gave a criterion for equicontinuity of a family $\mathfrak{F}$ of K-q.c. mappings in the spherical metric. It is a consequence of the Arzéla-Ascoli theorem that equicontinuous families are normal (i.e., each sequence has a locally uniformly convergent subsequence). It follows from Property 12.6 that limits of convergent sequences are again K-q.c. or constant. As an example:

<u>PROPERTY 12.7</u>. Let $\mathfrak{F}$ be the family of all K-q.c. mappings f of a domain D such that $\bar{\mathbb{C}}-f(D)$ contains two points (depending on f) with distance at least $d>0$ (independent of f) apart. Then $\mathfrak{F}$ is normal in the spherical metric.

In the following chapter we shall present a variational method for attacking extremal problems over families of q.c. mappings. We shall give the method even in the general situation of families of q.c. mappings f where the dilatation quotient D_f is bounded by a function of z . For that reason we make the following definition.

<u>DEFINITION</u>. Suppose K(z) is a measurable function on a domain D , with $\|K(z)\|_\infty < \infty$. If f is a $\|K(z)\|_\infty$ - q.c. mapping of D with $D_f(z) \le K(z)$ a.e., we shall call f a <u>K(z)-q.c. mapping</u>.

It is a result of K. Strebel [S20] that Property 12.6 is true for sequences of K(z)-q.c. mappings:

PROPERTY 12.6'. If f_n is a sequence of $K(z)$-q.c. mappings of a domain D, converging locally uniformly to f, then f is again a $K(z)$-q.c. mapping or a constant.

CHAPTER 13. A variational method for q.c. mappings

Variational methods for q.c. mappings were first used by P. P. Belinskiǐ [B1] and have been applied by P. A. Biluta, S. L. Kruškal', R. Kühnau, and others (see, e.g., [B3-4, K7-12] and the references there). In 1966 M. Schiffer [S10] gave a method that applied to families of continuously differentiable q.c. mappings. For modifications of Schiffer's approach, see Schiffer and Schober [S11] and H. Renelt [R1]. We shall give a blend of these approaches in a linear space framework.

Although we allow q.c. mappings to assume ∞, we shall avoid this value in linear space considerations. If f is a q.c. mapping of a domain D and E is a compact subset of $\hat{D} = D - f^{-1}(\{\infty\})$, then f belongs to $C(E)$, the Banach space of continuous functions on E with the supremum norm. (We shall not distinguish between a function whose domain contains E and the restriction of the function to E.) By the Riesz representation theorem, each $L \in C'(E)$, the topological dual of $C(E)$, has the representation

$$L(g) = \int_E g \, d\mu$$

where μ is a finite complex Borel measure supported in E. As before, we call μ a representing measure of $L \in C'(E)$ and use this measure to extend the definition of L to functions whose restriction to the support of μ is integrable with respect to μ. We denote

$$C_c'(\hat{D}) = \bigcup_{E \subset \hat{D}} C'(E).$$

We shall be concerned with functionals defined on families of q.c. mappings.

DEFINITION. Let $\mathfrak{F}$ be a family of q.c. mappings of a domain D and $f \in \mathfrak{F}$. We shall say that a real functional λ defined on $\mathfrak{F}$ has a complex Gâteaux derivative at f relative to $\mathfrak{F}$ if an $L \in C_c'(D-f^{-1}(\{\infty\}))$ and representing measure μ for L exist such that

$$\lambda(f^*) = \lambda(f) + \varepsilon \,\mathrm{Re}\, L(h) + o(\varepsilon)$$

whenever $f^* \in \mathfrak{F}$, $\varepsilon > 0$, $f^* = f + \varepsilon h + o(\varepsilon)$, and $(f^*)^{-1}(\{\infty\})$ does not belong to the support of μ.

Denote by χ_E the characteristic function of the set E, i.e.,

$$\chi_E(w) = \begin{cases} 1 & \text{if } w \in E \\ 0 & \text{if } w \notin E. \end{cases}$$

We begin by constructing variations:

LEMMA 13.1. Let E be a compact set and $a(w)$ a measurable function with $|a| = \chi_E$. Then for each ε, $0 < \varepsilon < 1$, there exists a $\frac{1+\varepsilon}{1-\varepsilon}$ - q.c. mapping Φ of the plane, that satisfies the Beltrami equation

$$\Phi_{\overline{w}} = \varepsilon a \Phi_w$$

and has normalization $\Phi(w) = w + o(1)$ near ∞. Moreover, Φ has the representation

$$\Phi(w) = w - \frac{\varepsilon}{\pi} \iint_E \frac{a(\zeta)}{\zeta - w}\, d\xi\, d\eta + O(\varepsilon^2)$$

where $O(\varepsilon^2)/\varepsilon^2$ is uniformly bounded depending only on E.

Proof. The existence and uniqueness of Φ are consequences of Proposition 12.1. By Proposition 12.2 we may represent

$$\Phi(w) = w + \varepsilon C(I - \varepsilon a \mathcal{H})^{-1} a(w).$$

Fix a p, $1 < p < 2$, and let $(1/p) + (1/q) = 1$. Assume

$0 < \varepsilon \le 1/(2\|\aleph\|_q)$. Then

$$\begin{aligned} |\Phi(w) - w - \varepsilon Ca(w)| &= \varepsilon |C[(I-\varepsilon a\aleph)^{-1} - I]a(w)| \\ &= \varepsilon^2 |Ca\aleph(I-\varepsilon a\aleph)^{-1}a(w)| \\ &\le \frac{\varepsilon^2}{\pi} \|a(\zeta)/(\zeta-w)\|_p \|\aleph(I-\varepsilon a\aleph)^{-1}a\|_q \\ &\le \frac{\varepsilon^2}{\pi} \Big(\iint_E \frac{d\xi d\eta}{|\zeta-w|^p}\Big)^{1/p} \cdot 2\|\aleph\|_q (m(E))^{1/q} . \end{aligned}$$

This estimate is uniformly bounded in w . Since p is fixed, there is a bound for the coefficient of ε^2 that depends only on E .

<u>DEFINITION</u>. Let f belong to a family $\mathfrak{F}$ of $K(z)$-q.c. mappings of a domain D , and set $D_1 = \{z \in D : K(z) > 1\}$. We shall say that <u>variations of f can be normalized in $\mathfrak{F}$</u> if, for each variation Φ with the support of a contained in $f(D_1)$ and with $D_{\Phi\circ f}(z) \le K(z)$ a.e., there exists a Möbius transformation τ such that $\tau \circ \Phi \circ f \in \mathfrak{F}$. Furthermore, we require that there exists a function $\Psi(\zeta,w)$ (<u>variational derivative relative to $\mathfrak{F}$ of f at ζ</u>) such that

$$\tau \circ \Phi(w) = w + \frac{\varepsilon}{\pi} \iint_{f(D_1)} a(\zeta)\Psi(\zeta,w)\, d\xi d\eta + o(\varepsilon) , \qquad w \in f(D) ,$$

where $o(\varepsilon)/\varepsilon \to 0$ locally uniformly on $f(D)$ as $\varepsilon \to 0$.

If $L \in C_c'(D - f^{-1}(\{\infty\}))$ has a representing measure μ such that $\Psi(\zeta, f(z))$ is locally integrable on $f(D_1) \times D$ with respect to $(d\xi d\eta) \times d\mu$ and

$$L(\Psi(\zeta,f)) \equiv \iint_D \Psi(\zeta, f(z))\, d\mu(z) \ne 0 \quad \text{for a.e. } \zeta \in f(D_1) ,$$

we shall say that <u>$L(\Psi(\cdot,f))$ is essentially nonzero</u>.

The above conditions assure that desired variations of f still belong to the family $\mathfrak{F}$. We shall study some families and functionals with this property in Chapter 14. We are now prepared for the principal theorem of the variational method.

THEOREM 13.2. Let f belong to a family $\mathfrak{F}$ of $K(z)$-q.c. mappings of a domain D. Suppose that f maximizes a real functional λ defined on $\mathfrak{F}$, i.e.,

$$\lambda(f) = \max_{\mathfrak{F}} \lambda .$$

Assume that variations of f can be normalized in $\mathfrak{F}$ and that λ has a Gâteaux derivative L at f relative to $\mathfrak{F}$ such that $L(\Psi(\cdot,f))$ is essentially nonzero.

Then f satisfies the differential equation

$$\text{(A)} \qquad f_{\bar{z}}(z) = k(z)\frac{|L(\Psi(f(z),f))|}{L(\Psi(f(z),f))}\overline{f_z(z)} \qquad \text{a.e. in } D,$$

where $k(z) = [K(z)-1]/[K(z)+1]$.

REMARK. We define the right side of the differential equation (A) to be zero whenever $k(z) = 0$. Then it is equivalent to the statement that

$$D_f(z) = K(z) \qquad \text{a.e. in } D$$

and

$$\arg \nu_f(z) = -\arg L(\Psi(f(z),f)) \quad \text{a.e. in } D_1 = \{z \in D : k(z) > 0\}.$$

That is, f has maximum distortion a.e., and its direction in $f(D_1)$ is $-\tfrac{1}{2}\arg L(\Psi(w,f))$ a.e. It follows that the direction of maximum distortion in $f(D_1)$ occurs a.e. along trajectories, when they exist, of the differential equation

$$L(\Psi(w,f))(dw)^2 > 0 .$$

Proof of Theorem 13.2. Let

$$\Sigma_\delta = \{z \in D : D_f(z) < K(z) - \delta\} ,$$

and suppose that Σ_δ has positive measure for some $\delta > 0$. Then $\Sigma_\delta \subset D_1$, and $f(\Sigma_\delta)$ has positive measure by Property 12.4. Let E be any compact subset of $f(\Sigma_\delta)$, and let Φ be the variation function of Lemma 13.1 with

$$a(w) = e^{i\alpha}\chi_E(w) , \qquad \alpha \text{ real} .$$

Then for all $\varepsilon \le \delta/[2\|K(z)\|_\infty - \delta]$ we have $D_{\Phi\circ f}(z) \le K(z)$ a.e., so that $f^* = \tau \circ \Phi \circ f \in \mathfrak{F}$ for appropriate Möbius transformations τ. Since f is an extremal function, it follows that

$$0 \ge \lambda(f^*) - \lambda(f) = \varepsilon\, \mathrm{Re}\{L(\frac{1}{\pi}\iint_E e^{i\alpha}\Psi(\zeta,f)\,d\xi d\eta)\} + o(\varepsilon) .$$

Divide by ε, and let $\varepsilon \to 0$. Then

$$\mathrm{Re}\{e^{i\alpha}L(\iint_E \Psi(\zeta,f)\,d\xi d\eta)\} \le 0 .$$

Since this is valid for all real α, we have

$$L(\iint_E \Psi(\zeta,f)\,d\xi d\eta) = 0 .$$

By Fubini's theorem we may interchange L and $\iint_E$. Therefore, since E is arbitrary, we must have $L(\Psi(\zeta,f)) = 0$ for a.e. $\zeta \in f(\Sigma_\delta)$. Since we assumed that $L(\Psi(\cdot,f))$ is essentially nonzero, we conclude that Σ_δ must have measure zero for all $\delta > 0$. As a consequence, $D_f(z) = K(z)$ a.e. in D.

Now let

$$p(z) = \sqrt{f_z(z)f_{\bar z}(z)L(\Psi(f(z),f))}$$

where we choose values of the square root with nonnegative imaginary part, and define

$$T_\delta = \{z \in D_1 : L(\Psi(f(z),f)) \ne 0 \text{ and } k(z)\cdot \mathrm{Im}\{p(z)/|p(z)|\} > \delta\} .$$

Suppose that T_δ has positive measure for some $\delta > 0$. As before, $f(T_\delta)$ then has positive measure, and we let E be an arbitrary compact subset of $f(T_\delta)$. Let Φ be the variation function of Lemma 13.1 with

$$a(w) = -i\frac{p(z)\,\overline{|L(\Psi(w,f))|}}{|p(z)|\;L(\Psi(w,f))}\chi_E(w) \qquad z = f^{-1}(w) .$$

Then $\Re e\{\bar{a} f_z f_{\bar{z}}\} = -|f_z f_{\bar{z}}|\,\mathcal{I}m\{p/|p|\}$ and $|f_{\bar{z}}| = k|f_z|$ a.e. in $f^{-1}(E)$, so that

$$\left|\frac{(\Phi\circ f)_{\bar{z}}}{(\Phi\circ f)_z}\right|^2 = \frac{|f_{\bar{z}} + \epsilon a\overline{f_z}|^2}{|f_z + \epsilon a\overline{f_{\bar{z}}}|^2} = \frac{|f_{\bar{z}}|^2 + \epsilon^2|f_z|^2 - 2\epsilon|f_z f_{\bar{z}}|\,\mathcal{I}m\{p/|p|\}}{|f_z|^2 + \epsilon^2|f_{\bar{z}}|^2 - 2\epsilon|f_z f_{\bar{z}}|\,\mathcal{I}m\{p/|p|\}}$$

$$= \frac{k^2 + \epsilon^2 - 2\epsilon k\,\mathcal{I}m\{p/|p|\}}{1 + \epsilon^2 k^2 - 2\epsilon k\,\mathcal{I}m\{p/|p|\}} \le \frac{k^2 + \epsilon^2 - 2\epsilon\delta}{1 + \epsilon^2 k^2 - 2\epsilon\delta}$$

by monotonicity. By adding $\epsilon^2(1-k^2)^2$ to the last numerator one easily obtains the bound k^2 as long as $\epsilon \le \delta$. Consequently, $D_{\Phi\circ f}(z) \le K(z)$ a.e. in $f^{-1}(E)$. Since Φ is conformal off E, we have $D_{\Phi\circ f}(z) \le K(z)$ a.e. in D for $\epsilon \le \delta$. So $f^* = \tau\circ\Phi\circ f \in \mathfrak{F}$ for appropriate Möbius transformations τ, and we conclude as before that

$$0 \ge \lambda(f^*) - \lambda(f) = \epsilon\Re e\{L(\tfrac{1}{\pi}\iint_E a(\zeta)\Psi(\zeta,f)\,d\xi d\eta)\} + o(\epsilon) .$$

Divide by ϵ, let $\epsilon \to 0$, and use Fubini's theorem. Then

$$0 \ge \iint_E \Re e\{a(\zeta)L(\Psi(\zeta,f))\}\,d\xi d\eta = \iint_E \mathcal{I}m\{p/|p|\}\,|L|\,d\xi d\eta \ge \delta\iint_E |L|/k\,d\xi d\eta .$$

Consequently, $L(\Psi(\cdot,f)) = 0$ a.e. in E, and since E is arbitrary, a.e. in $f(T_\delta)$. Since we assumed that $L(\Psi(\cdot,f))$ is essentially nonzero, we can conclude as before that T_δ has measure zero for all $\delta > 0$. Since $k > 0$ and $L(\Psi(\cdot,f)) \ne 0$ a.e. in D_1, it follows that $\mathcal{I}m\{p/|p|\} = 0$ and $p^2 > 0$ a.e. in D_1. As a consequence,

$$\arg \nu_f(z) = -\arg L(\Psi(f(z),f)) \qquad \text{a.e. in } D_1 ,$$

and the proof is complete.

COROLLARY 13.3. In addition to the hypotheses of Theorem 13.2, suppose that the function $L(\Psi(\cdot,f))$ is analytic and does not vanish in a neighborhood of a finite point $w_o = f(z_o)$. Then f satisfies

$$(B) \qquad (J \circ f)_{\bar{z}} = k(z)\overline{(J \circ f)_z}$$

a.e. in a neighborhood of z_o , where J denotes any local integral

$$J(w) = \int^{w} \sqrt{L(\Psi(\zeta,f))}\, d\zeta \; .$$

Moreover, if $K(z)$ has in a neighborhood of z_o partial derivatives of order n , $n \geq 0$, that satisfy a Hölder condition, then in a neighborhood of z_o , f has partial derivatives of order $n+1$ that satisfy a Hölder condition, and equations (A) and (B) are satisfied identically.

Proof. Since

$$(J \circ f)_{\bar{z}} = \sqrt{L}\, f_{\bar{z}} = \sqrt{L}\, k\, \frac{|L|}{L}\, \overline{f_z} = k\overline{\sqrt{L}\, f_z} = k\overline{(J \circ f)_z}\, ,$$

equation (B) is an immediate consequence of (A).

By Property 12.5, q.c. mappings are locally Hölder continuous. For $n=0$ we are assuming that K itself is Hölder continuous. Then $\nu_f = k|L(\Psi(\cdot,f))|/L(\Psi(\cdot,f))$ is Hölder continuous and by Property 12.5, f has first order partial derivatives that satisfy a Hölder condition. If now K also has partial derivatives of order $n=1$ that satisfy a Hölder condition, then ν_f does also and by Property 12.5, f has second order partial derivatives that satisfy a Hölder condition. For the general case one proceeds by finite induction. Since K, $f_{\bar{z}}$, f_z, and $L(\Psi(f(\cdot),f))$ are continuous, equations (A) and (B) become identities in a neighborhood of z_0.

COROLLARY 13.4. In addition to the hypotheses of Theorem 13.2, suppose that the function $L(\Psi(\cdot,f))$ is analytic and does not vanish in a neighborhood of a finite point $w_0 = f(z_0)$ and that $K(z)$ is constant in a neighborhood of z_0. Then f has partial derivatives of all orders, and

$$J \circ f - k\overline{J \circ f}$$

is analytic in a neighborhood of z_0. Here J denotes any local integral

$$J(w) = \int^{w} \sqrt{L(\Psi(\zeta,f))}\,d\zeta .$$

Proof. It follows from Corollary 13.3 that f has partial derivatives of all orders. Since k is constant, the differential equation (B) can be written as

$$[J \circ f - k\overline{J \circ f}]_{\bar{z}} = 0 .$$

Consequently, the function $J \circ f - k\overline{J \circ f}$ has partial derivatives of all orders and satisfies the Cauchy-Riemann equations; hence it is analytic.

REMARKS. A q.c. mapping f is called a Teichmüller mapping if it satisfies the Beltrami equation

$$f_{\bar{z}} = k\,\frac{|\varphi|}{\varphi}\,f_z$$

where k, $0<k<1$, is constant and φ is analytic. Under the hypotheses of Corollary 13.4, equation (A) implies that f^{-1} is a Teichmüller mapping in a neighborhood of w_0.

In solving problems one would like to know the extremal

functions f, or at least enough information about them to determine extreme values of functionals. The variational procedure of this chapter leads to information about f from the differential equation (A) and about $J \circ f$ from the differential equation (B). In practice we shall try to determine $J \circ f$ explicitly. For this purpose the analyticity properties of $J \circ f - k\overline{J \circ f}$ will be extremely important and will lead to interesting new boundary value problems.

Since conformal variations do not introduce any additional distortion, Schiffer's boundary variations (Appendix C) can often be used to give information about $\partial f(D)$ for an extremal function f. Under appropriate conditions, it will usually be to the effect that boundary components are piecewise analytic arcs satisfying the differential equation $L(\Psi(w,f))(dw)^2 > 0$ (cf. p. 141).

CHAPTER 14. Application to families of conformal and q.c. mappings

We shall give some applications to the following families of functions:

$S_K = \{f : f$ is a K-q.c. mapping of $\mathbb{C}$ into $\bar{\mathbb{C}}$, and $f|_U \in S\}$

$\Sigma_K = \{g : g$ is a K-q.c. mapping of $\mathbb{C}$ onto $\mathbb{C}$, and $g|_{|z|>1} \in \Sigma\}$.

Functions in S_K may assume ∞ in $|z|>1$, while functions in Σ_K leave ∞ fixed. We shall also consider the hyperbolic families

$S_{K,R} = \{f : f$ is a K-q.c. mapping of $|z|<R$ into $\bar{\mathbb{C}}$, and $f|_U \in S\}$

$\Sigma_{K,r} = \{g : g$ is a K-q.c. mapping of $|z|>r$ into $\mathbb{C}$, and $g|_{|z|>1} \in \Sigma\}$

for $1<R<\infty$ and $0<r<1$. We denote by Σ'_K and $\Sigma'_{K,r}$ those subsets of Σ_K and $\Sigma_{K,r}$, respectively, for which $g(z) = z + o(1)$ as $z \to \infty$, i.e., for which $g|_{|z|>1} \in \Sigma'$.

When the meaning is clear, we shall not distinguish between, say, a function in S_K and its restriction, which is in S. Then we may view S_K, Σ_K, and Σ'_K as those subsets of S, Σ, and Σ', respectively, of functions that have K-q.c. extensions to the Riemann sphere.

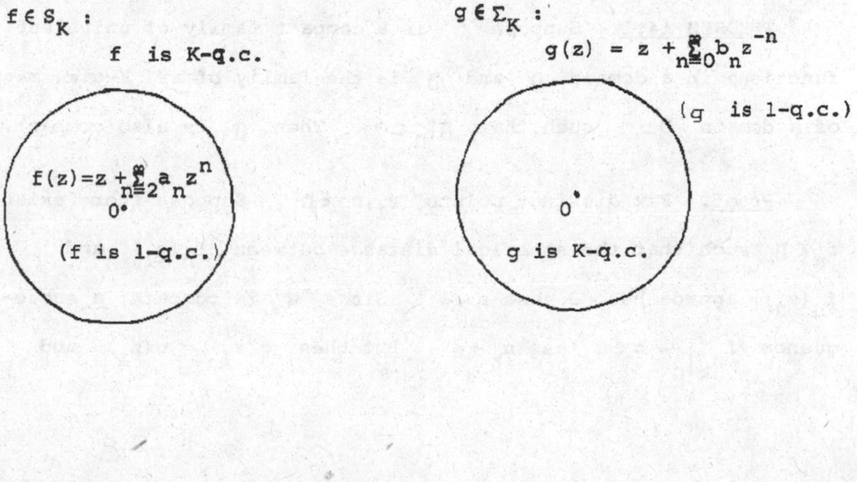

S_1 contains only the Möbius transformations $z/(1-\eta z)$, $|\eta| \le 1$, and Σ_1 only contains translations of the identity. On the other extreme, the classes S_K and Σ_K (restricted appropriately) are dense in S and Σ, respectively, as $K \to \infty$. Therefore the parameter K provides an interpolation between elementary Möbius transformations and the full classes S and Σ. We shall assume henceforth that $1 < K < \infty$.

The classes $S_{K,R}$, $\Sigma_{K,r}$, and $\Sigma'_{K,r}$ have been introduced because their functions map onto hyperbolic domains. The presence of a nontrivial boundary adds a new dimension to the solution of extremal problems, that incorporates Schiffer's boundary variations (Appendix C). Properly restricted, the classes $S_{K,R}$ and $\Sigma_{K,r}$ are dense in S and Σ, respectively, both as $K \to \infty$ and as $R, r \to 1$. Other relations, such as $S_K \subset S_{K,R}$ and $\Sigma_K \subset \Sigma_{K,r}$, are evident.

<u>PROBLEM</u>. What is the smallest $\hat{K} = \hat{K}(K,R)$ such that if $f \in S$ has a K-q.c. extension to $|z| < R$, then f has a $\hat{K}$-q.c. extension to all of $\mathbb{C}$? That is, characterize $\hat{K}(K,R) = \inf\{K' : S_{K,R}|_U \subset S_{K'}|_U\}$.

<u>THEOREM 14.1</u>. Suppose $\mathfrak{G}$ is a compact family of univalent functions in a domain Ω and $\mathfrak{F}$ is the family of all K-q.c. mappings of a domain $D \supset \Omega$ such that $\mathfrak{F}|_\Omega \subset \mathfrak{G}$. Then $\mathfrak{F}$ is also compact.

<u>Proof</u>. Fix distinct points $z_1, z_2 \in \Omega$. Suppose there exist $f_n \in \mathfrak{F}$ such that the spherical distance between $f_n(z_1)$ and $f_n(z_2)$ approaches 0 as $n \to \infty$. Since $\mathfrak{G}$ is compact, a subsequence $f_{n_k}|_\Omega \to g \in \mathfrak{G}$ as $n_k \to \infty$. But then $g(z_1) = g(z_2)$ and

contradicts the univalence of g. Therefore each member of the family $\mathfrak{J}|_{D-\{z_1,z_2\}}$ omits values at a spherical distance at least d for some fixed $d > 0$. By Property 12.7 this family is normal in the spherical metric, and by the maximum principle (follows since the mappings are open) so is $\mathfrak{J}$. Therefore by Property 12.6, $\mathfrak{J}$ is compact; constant limits are eliminated since $\mathfrak{G}$ is a univalent family.

Since S and Σ' are compact families of univalent functions, the following is a special case of Theorem 14.1:

<u>COROLLARY 14.2</u>. S_K, Σ'_K, $S_{K,R}$, $\Sigma'_{K,r}$ are compact families.

<u>THEOREM 14.3</u>. Let $f \in S_K$ $(S_{K,R})$. Then variations of f can be normalized in S_K $(S_{K,R})$, and the variational derivative relative to S_K $(S_{K,R})$ of f at ζ is

$$\Psi(\zeta,w) = \frac{w^2}{\zeta^2(w-\zeta)}.$$

<u>Proof</u>. For each variation Φ with the support of a in $f(|z|>1)$ and with $D_{\Phi\circ f}(z) \le K$ a.e., one has $\tau\circ\Phi\circ f \in S_K$ where

$$\tau\circ\Phi(w) = [\Phi(w)-\Phi(0)]/\Phi'(0) = w + \frac{\varepsilon}{\pi}\iint_{f(|z|>1)} a(\zeta)w^2/[\zeta^2(w-\zeta)]d\xi d\eta + O(\varepsilon^2)$$

and $O(\varepsilon^2)/\varepsilon^2$ has a uniform bound on compact sets as $\varepsilon \to 0$. The variational derivative relative to S_K at ζ is therefore $\Psi(\zeta,w) = w^2/[\zeta^2(w-\zeta)]$ for any $f \in S_K$. The proof for $S_{K,R}$ is identical.

<u>THEOREM 14.4</u>. Let $g \in \Sigma_K$ $(\Sigma'_K, \Sigma_{K,r}, \Sigma'_{K,r})$. Then variations of g can be normalized in Σ_K $(\Sigma'_K, \Sigma_{K,r}, \Sigma'_{K,r})$ and the variational derivative relative to Σ_K $(\Sigma'_K, \Sigma_{K,r}, \Sigma'_{K,r})$ of g

at ζ is

$$\Psi(\zeta, w) = 1/(w-\zeta) .$$

<u>Proof</u>. The variations Φ, themselves, preserve the normalization of Σ_K, Σ_K', $\Sigma_{K,r}$, and $\Sigma_{K,r}'$.

Since restrictions of functions in S_K $(S_{K,R})$ and Σ_K $(\Sigma_{K,r})$ belong to S and Σ, respectively, every extremal problem for S and Σ can now be directed to S_K $(S_{K,R})$ and Σ_K $(\Sigma_{K,r})$. For example, if λ is a real Gâteaux differentiable functional defined on Σ, then λ may be defined for $g \in \Sigma_K$ as $\lambda(g|_{|z|>1})$. The Gâteaux derivative L of λ at $g|_{|z|>1}$ has a representing measure μ with compact support E in $|z| > 1$. We may then use

$$L(h) = \int_E h \, d\mu$$

to extend the definition of L to Σ_K (in fact, to all functions integrable over E with respect to μ). In this way, the extended λ has the extended L as its Gâteaux derivative. However, in order to apply the variational method we shall need only

$$L(\Psi(\zeta, g)) = L(1/(g-\zeta)) = \int_E 1/(g-\zeta) d\mu \quad \text{for} \quad |\zeta| \le 1 .$$

We are now prepared for our first elementary application of the variational method:

<u>THEOREM 14.5</u> (Kühnau [K8]). If $g \in \Sigma_K$ and $g(z) = z + \sum_{n=0}^{\infty} b_n z^{-n}$ for $|z| > 1$, then

$$|b_1| \le k = (K-1)/(K+1) .$$

Equality occurs iff

$$g(z) = \begin{cases} z + b_0 + ke^{i\alpha}/z & \text{for } |z| > 1 \\ z + b_0 + ke^{i\alpha}\bar{z} & \text{for } |z| \le 1 . \end{cases}$$

Proof. Since Σ_K' is compact, an extremal function g exists for the problem $\max_{\Sigma_K'} \Re e\, b_1 = \max_{\Sigma_K} \Re e\, b_1$. The functional $\Re e\, b_1$ has complex Gâteaux derivative $L(h) = b_1$, so that $L(1/(g-\zeta)) = 1$ and $J(w) = \int^w \sqrt{1}\, d\zeta = w$. Therefore Corollary 13.4 implies that $J \circ g - k\overline{J \circ g} = g - k\bar{g}$ is analytic in $|z| < 1$. Consequently,

$$0 = \frac{1}{2\pi i} \int_{|z|=1} [g(z) - k\overline{g(z)}] z^{n-1} dz = \begin{cases} b_1 - k & \text{for } n = 1 \\ b_n & \text{for } n > 1 . \end{cases}$$

That is, in the extreme case $b_1 = k$ and $b_n = 0$ for $n > 1$. The inequality for the modulus of b_1 follows from that for the real part through the familiar rotation $g(z) \to e^{i\alpha} g(e^{-i\alpha} z)$.

Note the ease with which the analyticity statement of Corollary 13.4 led both to the extreme value of the functional and to the extremal function in the above proof. At first glance, then, the variational method for q.c. mappings appears to be a very powerful tool for solving extremal problems.

Observe that the bound of Theorem 14.5 agrees with the classical result (Theorem B1) as $K \to \infty$. We note some of the standard consequences of such an estimate in the following corollaries:

COROLLARY 14.6 ([K8]). If $g \in \Sigma_K$, $g(0) = 0$, and $g(z) = z + \sum_{n=0}^{\infty} b_n z^{-n}$ for $|z| > 1$, then

$$|b_0| \le 2k , \quad k = (K-1)/(K+1) .$$

Equality occurs iff

$$g(z) = \begin{cases} z[1 + ke^{i\alpha}/z]^2 & \text{for } |z| > 1 \\ z + 2ke^{i\alpha}|z| + k^2e^{2i\alpha}\bar{z} & \text{for } |z| \le 1 . \end{cases}$$

<u>Proof</u>. Apply Theorem 14.5 to the odd function $\sqrt{g(z^2)}$, which is again in Σ_K .

<u>COROLLARY 14.7</u> ([K8]). Suppose $f \in S_K$ and $f(z) = z + \sum_{n=2}^{\infty} a_n z^n$ for $z \in U$. Then

$$|a_3 - a_2^2| \le k .$$

If, in addition, $f(\infty) = \infty$, then

$$|a_2| \le 2k$$

and $|a_2| = 2k$ iff

$$f(z) = \begin{cases} z/(1 + ke^{i\alpha}z)^2 & \text{for } |z| < 1 \\ |z|^2/[\bar{z} + 2ke^{i\alpha}|z| + k^2e^{2i\alpha}z] & \text{for } |z| \ge 1 . \end{cases}$$

<u>Proof</u>. Apply Theorem 14.5 and Corollary 14.6 to $1/f(1/z) \in \Sigma_K$.

We now turn to similar problems for the classes $\Sigma_{K,r}$ and $S_{K,R}$. The presence of a boundary requires more information, which will be supplied by Schiffer's fundamental lemma (Appendix C).

<u>THEOREM 14.8</u> (McLeavey [M2]) . If $g \in \Sigma_{K,r}$ and $g(z) = z + \sum_{n=0}^{\infty} b_n z^{-n}$ for $|z| > 1$, then

$$|b_1| \le (k + r^2)/(1 + kr^2) .$$

Equality occurs iff

$$g(z) = \begin{cases} z + [(k+r^2)/(1+kr^2)]e^{i\alpha}/z + b_0 & \text{for } |z| > 1 \\ [1/(1+kr^2)][z + r^2e^{i\alpha}/z + ke^{i\alpha}\bar{z} + kr^2/\bar{z}] + b_0 & \text{for } r < |z| \le 1 . \end{cases}$$

<u>Proof</u>. After a translation and rotation, it is sufficient to consider the problem $\max_{\Sigma'_{K,r}} \operatorname{Re} b_1$. As in the proof of Theorem 14.5, an extremal function g has the property that $g-k\bar{g}$ is analytic in $r<|z|<1$. Since variations of the continuum $\gamma = \mathbb{C}-g(|z|>r)$ as in Theorem C.3 are conformal and properly normalized, they operate within the family $\Sigma'_{K,r}$. Therefore, by applying Schiffer's fundamental lemma as in the proof of Theorem 10.13, we conclude that γ is an analytic arc satisfying the differential equation $(dw)^2 > 0$. That is, γ is a horizontal segment. The mapping $W = w-k\bar{w}$ takes γ onto another horizontal segment. Therefore the function $g-k\bar{g}$ can be continued analytically across $|z| = r$, and the continuation satisfies the identity

$$g(z) - k\overline{g(z)} = \overline{g(r^2/\bar{z})} - k\, g(r^2/\bar{z}) + \text{imaginary constant}$$

in $r^2 \le |z| \le 1$. We now compute

$$\left.\begin{array}{ll} b_n & \text{for } n>1 \\ b_1-k & \text{for } n=1 \end{array}\right\} = \frac{1}{2\pi i}\int_{|z|=1}[g(z)-k\overline{g(z)}]z^{n-1}dz = \frac{1}{2\pi i}\int_{|z|=r^2}[g(z)-k\overline{g(z)}]z^{n-1}dz$$

$$= \frac{1}{2\pi i}\int_{|z|=r^2}[\overline{g(r^2/\bar{z})}-kg(r^2/\bar{z})]z^{n-1}dz = \frac{r^{2n}}{2\pi i}\int_{|z|=1}[\overline{g(z)}-kg(z)]z^{n-1}dz$$

$$= \begin{cases} -r^{2n}kb_n & \text{for } n>1 \\ r^2(1-kb_1) & \text{for } n=1 . \end{cases}$$

Since $|r^{2n}k| < 1$, we conclude that $b_n = 0$ for $n>1$ and $b_1 = (k+r^2)/(1+kr^2)$. That is, the extremal function

$$g(z) = z + [(k+r^2)/(1+kr^2)]/z \quad \text{for } |z|>1 .$$

Consequently, for $|z| = 1$

$$g(z) - k\overline{g(z)} = [(1-k^2)/(1+kr^2)][z + r^2/z] ,$$

and therefore for $r < |z| < 1$ also. Thus

$$g(z) = [1/(1+kr^2)][z + r^2/z + k\bar{z} + kr^2/\bar{z}] \quad \text{for } r < |z| \le 1 .$$

EXERCISE. Let D be a domain containing a neighborhood of ∞ and $\hat{D}$ a domain containing D. Let $\Sigma'_{K,\hat{D}}(D)$ be the class of all K-q.c. mappings g of $\hat{D}$ that are conformal in D and have the normalization $g(z) = z + \sum_{n=1}^{\infty} b_n z^{-n}$ at ∞. By considering the problem $\max_{\Sigma'_{K,\hat{D}}(D)} \Re\{e^{-2i\alpha} b_1\}$, show there exists a $g \in \Sigma'_{K,\hat{D}}(D)$ such that $\mathbb{C}-g(D)$ consists entirely of points and line segments of inclination α, $D_g \equiv K$ in $\hat{D}-D$, and α is the direction of maximum distortion in $g(\hat{D}-D)$.

COROLLARY 14.9 ([M2]). If $g \in \Sigma_{K,r}$, $g \ne 0$, and $g(z) = z + \sum_{n=0}^{\infty} b_n z^{-n}$ for $|z| > 1$, then

$$|b_0| \le 2(k+r)/(1+kr) .$$

Proof. Apply Theorem 14.8 to $\sqrt{g(z^2)} \in \Sigma_{K,\sqrt{r}}$.

COROLLARY 14.10 ([M2]). Suppose $f \in S_{K,R}$ and $f(z) = z + \sum_{n=2}^{\infty} a_n z^n$ for $z \in U$. Then

$$|a_3 - a_2^2| \le (1 + kR^2)/(k + R^2) .$$

If, in addition, $f \ne \infty$, then

$$|a_2| \le 2(1+kR)/(k+R) .$$

Proof. Apply Theorem 14.8 and Corollary 14.9 to $1/f(1/z) \in \Sigma_{K,1/R}$.

EXERCISE. Find the extremal functions in Corollaries 14.9 and 14.10.

The article of J. McLeavey [M2] contains the above results as a special case. It concerns, more generally, extremal problems for those subclasses of S and Σ of functions that have $K(z)$-q.c. extensions where $K(z) = K(|z|)$.

We shall now consider general linear problems for the class S_K, considered as a subset of S. As before, when the meaning is clear, we shall not distinguish between a function in S_K and its restriction to U, which is in S.

THEOREM 14.11. Suppose $f \in S_K$, $L \in H'(U)$ is not of the form $L(h) = \alpha h(0) + \beta h'(0)$, and $\Re L(f) = \max_{S_K} \Re L$. Then

(i) $\Re L(f^2/(f-w)) > 0$ for $w \in f(1 < |z| \le \infty)$, and

(ii) $\infty \in f(|z| = 1)$.

Proof. If $w \in \bar{C} - f(U)$, then $f^* = f/(1-\frac{1}{w}f) \in S_K$. Therefore $\Re L(f^*) \le \Re L(f)$, and so $\Re L(f^2/(f-w)) \ge 0$. The function $L(f^2/(f-w))$ is analytic for $w \in f(1 < |z| \le \infty)$ and not identically zero by Lemma 9.4. Therefore (i) follows from the minimum principle. Since $L(f^2/(f-w))$ vanishes for $w = \infty$, part (ii) is a consequence of (i).

EXERCISE (Schiffer and Schober [S12]). Suppose $f \in S_K$, $L \in H'(U)$ is not of the form $L(h) = \alpha h(0) + \beta h'(0)$, and $|L(f)| = \max_{S_K} |L|$. Show that (i) $L(f^2/(f-w)) \ne 0$ for $w \in f(1 < |z| \le \infty)$, (ii) $\infty \in f(|z|=1)$, and (iii) $f(1 < |z| \le \infty)$

lies in the region $\{w : \Re e[L(f)/L(f^2/(f-w))] > \frac{1}{2}\}$.

We single out the special case of the coefficient problem for S_K:

<u>COROLLARY 14.12</u>. Suppose f is an extremal function for the problem $\max_{S_K} \Re e\, a_n$, $n \geq 2$. Then

(i) $\Re e \sum_{m=2}^{n} \frac{a_n^{(m)}}{w^{m-1}} < 0$ for $w \in f(1 < |z| \leq \infty)$ where

$$[f(z)]^m = \sum_{n=m}^{\infty} a_n^{(m)} z^n \quad \text{for } z \in U \text{, and}$$

(ii) $\infty \in f(|z| = 1)$.

Theorem 14.11 has significant consequences for attacking linear problems by the variational method. Indeed, suppose f is an extremal function for the problem $\max_{S_K} \Re e\, L$, where $L \in H'(U)$ is nontrivial. Then by Theorem 14.11,

$$J(w) = \int^{w} \sqrt{L(f^2/(f-\zeta))}\, \frac{d\zeta}{\zeta}$$

defines a single-valued analytic function in $f(1 < |z| \leq \infty)$. Consequently, by Corollary 13.4 and Theorem 14.3 the function

$$b = J \circ f - k\overline{J \circ f}$$

is single valued and analytic in $|z| > 1$, and finite at ∞. In U the function

$$a = J \circ f$$

is locally analytic away from the support of a representing measure for L and from possible branch points arising from zeros of $L(f^2/(f - f(\cdot)))$. If we define

$$\tilde{b}(z) = \overline{b(1/\bar{z})} \qquad \text{for } z \in U \text{, and}$$

$$\tilde{a}(z) = \overline{a(1/\bar{z})} \qquad \text{for } |z| > 1 \text{,}$$

then on $|z| = 1$

$$b = a - k\bar{a} = a - k\tilde{a} \quad \text{and}$$

$$(1-k^2)a = b + k\bar{b} = b + k\tilde{b} \ .$$

We may therefore use

$$a = b + k\tilde{a}$$

to continue a analytically into $|z| > 1$ and

$$b = (1-k^2)a - k\tilde{b}$$

to continue b analytically into U. Thus the problem becomes one of determining a and b from their singularity structure and global analytic character.

Because of Corollary 14.12 (ii), it is clear that Corollary 14.7 does not solve the second coefficient problem in S_K. We shall now solve this problem.

<u>THEOREM 14.13</u> (Schiffer and Schober [S12]). Let $f \in S_K$ and $f(z) = z + \sum_{n=2}^{\infty} a_n z^n$ for $z \in U$. Then

$$|a_2| \leq 2 - 4\varkappa^2$$

where $\varkappa = \frac{1}{\pi} \arccos k \in (0, \frac{1}{2}]$. Equality occurs only for the function

$$f(z) = \begin{cases} \dfrac{4z}{(1-z)^2}\left[\left(\dfrac{1+\sqrt{z}}{1-\sqrt{z}}\right)^{\varkappa} + \left(\dfrac{1-\sqrt{z}}{1+\sqrt{z}}\right)^{\varkappa}\right]^{-2} & |z|<1 \\ -4(1-k^2)\left\{\dfrac{1-z}{\sqrt{z}}\left[\left(\dfrac{\sqrt{z}+1}{\sqrt{z}-1}\right)^{\varkappa} - \left(\dfrac{\sqrt{z}-1}{\sqrt{z}+1}\right)^{\varkappa}\right] + k\,\overline{\dfrac{1-z}{\sqrt{z}}\left[\left(\dfrac{\sqrt{z}+1}{\sqrt{z}-1}\right)^{\varkappa} - \left(\dfrac{\sqrt{z}-1}{\sqrt{z}+1}\right)^{\varkappa}\right]}\right\}^{-2} & |z| \geq 1 \end{cases}$$

and its rotations $e^{-i\alpha} f(e^{i\alpha} z)$, $\alpha \in (0, 2\pi)$.

<u>Proof</u>. We may assume that f is an extremal function for the problem $\max_{S_K} \operatorname{Re} a_2$. Then the general case will follow from a rotation. Moreover, we may assume that $a_2 > 0$ for f. We shall

make use of the previous discussion.

Since $L(f^2/(f-\zeta)) = -1/\zeta$, the function

$$J(w) = \int^w \sqrt{-1/\zeta}\,\frac{d\zeta}{\zeta} = \frac{2}{i\sqrt{w}}$$

and

$$a = J \circ f = \frac{2}{i\sqrt{f}} \qquad \text{for } z \in U,$$

$$b = J \circ f - k\,\overline{J \circ f} = \frac{2}{i}\left[\frac{1}{\sqrt{f}} + \frac{k}{\overline{\sqrt{f}}}\right] \qquad \text{for } |z| > 1.$$

Moreover, the function b is single valued and analytic for $|z| > 1$, and finite at ∞. In U the function a is double valued and analytic except at the origin.

If we define $\tilde{b}(z) = \overline{b(1/\bar{z})}$ as above, then

$$c = (1-k^2)a^2 + \tilde{b}^2$$

is single valued and analytic in U, except for a simple pole at the origin with residue $-4(1-k^2)$. On $|z| = 1$,

$$c = (1-k^2)a^2 + (\bar{a} - ka)^2 = 2\,\mathrm{Re}(a^2) - 2k|a|^2$$

is real. Consequently,

$$c(z) + 4(1-k^2)(z + 1/z)$$

is constant in U.

Let $f(z_o) = \infty$. Then $|z_o| = 1$ by Corollary 14.12 (ii), and $c(z_o) = 0$. Therefore

$$c(z) = 4(1-k^2)(2\,\mathrm{Re}\, z_o - z - 1/z) \qquad \text{for } z \in U.$$

We also compute

$$[\tilde{b}(0)]^2 = \lim_{z \to 0} [c(z) - (1-k^2)a(z)^2] = 4(1-k^2)(2\,\mathrm{Re}\, z_o - a_2).$$

Consider now

$$d = z^2(a\tilde{b}' - \tilde{b}a')^2 = z^2[a^2(\tilde{b}')^2 - \tfrac{1}{2}(a^2)'(\tilde{b}^2)' + \tilde{b}^2(a')^2] .$$

d is single valued and analytic in U, except for a simple pole at the origin with residue $-4(1-k^2)(2\,\Re e\, z_0 - a_2)$. On $|z| = 1$,

$$d(e^{i\theta}) = 4\Big[\Im m\,\Big\{\overline{a(e^{i\theta})\frac{d}{d\theta}a(e^{i\theta})}\Big\}\Big]^2 = \frac{16}{|f(e^{i\theta})|^2}\Big[\frac{d}{d\theta}\arg f(e^{i\theta})\Big]^2$$

is real. As before, $d + 4(1-k^2)(2\,\Re e\, z_0 - a_2)(z + 1/z)$ is constant in U. Since $d(z_0) = 0$,

$$d(z) = 4(1-k^2)(2\,\Re e\, z_0 - a_2)(2\,\Re e\, z_0 - z - 1/z) .$$

Set $A = \tilde{b} + i\sqrt{1-k^2}\,a$ and $B = \tilde{b} - i\sqrt{1-k^2}\,a$. Then

$$\frac{A'}{A} = \frac{h'}{2h} - i\sqrt{1-k^2}\,\frac{\sqrt{d}}{zh} \quad\text{and}\quad \frac{B'}{B} = \frac{h'}{2h} + i\sqrt{1-k^2}\,\frac{\sqrt{d}}{zh} .$$

Since $\lim_{z\to 0} A(z)/\sqrt{h(z)} = i$ and $\lim_{z\to 0} B(z)/\sqrt{h(z)} = -i$, one finds that

$$A(z) = i\sqrt{h(z)}\,\exp\{-i\sqrt{1-k^2}\int_0^z \frac{\sqrt{d(t)}}{th(t)}\,dt\} \quad\text{and}$$

$$B(z) = -i\sqrt{h(z)}\,\exp\{i\sqrt{1-k^2}\int_0^z \frac{\sqrt{d(t)}}{th(t)}\,dt\} .$$

Therefore,

$$\begin{aligned} a(z) &= [A(z) - B(z)]/[2i\sqrt{1-k^2}] \\ &= 2\sqrt{2\,\Re e\, z_0 - z - 1/z}\,\cosh\{\tfrac{1}{2}\sqrt{2\,\Re e\, z_0 - a_2}\int_0^z \frac{dt}{\sqrt{t(1-z_0t)(1-\bar{z}_0t)}}\} \end{aligned}$$

and

$$\begin{aligned} \tilde{b}(z) &= \tfrac{1}{2}[A(z) + B(z)] \\ &= -2i\sqrt{1-k^2}\sqrt{2\,\Re e\, z_0 - z - 1/z}\,\sinh\{\tfrac{1}{2}\sqrt{2\,\Re e\, z_0 - a_2}\int_0^z \frac{dt}{\sqrt{t(1-z_0t)(1-\bar{z}_0t)}}\} . \end{aligned}$$

So for $z \in U$,

$$\begin{aligned} f(z) &= -4/a(z)^2 \\ &= \frac{z}{(1-z_0z)(1-\bar{z}_0z)}\,\operatorname{sech}^2\{\tfrac{1}{2}\sqrt{2\,\Re e\, z_0 - a_2}\int_0^z \frac{dt}{\sqrt{t(1-z_0t)(1-\bar{z}_0t)}}\} . \end{aligned}$$

Evidently $f(z_0) = f(\bar{z}_0) = \infty$ in the above formula. Since f is one-to-one, this is possible only if $z_0 = \overline{z}_0 = \pm 1$. If $z_0 = \overline{z}_0 = -1$, then

$$f(1) = \tfrac{1}{4}\sec^2\{\tfrac{1}{4}\pi\sqrt{2+a_2}\} > 0 ,$$

in contradiction to Corollary 14.12 (i). Therefore $z_0 = \overline{z}_0 = 1$ and

$$f(z) = \frac{4z}{(1-z)^2}\left[\left(\frac{1+\sqrt{z}}{1-\sqrt{z}}\right)^{\sigma} + \left(\frac{1-\sqrt{z}}{1+\sqrt{z}}\right)^{\sigma}\right]^{-2} , \quad z \in U ,$$

where $\sigma = \tfrac{1}{2}\sqrt{2-a_2}$. For $|z| > 1$,

$$f(z) = -4(1-k^2)^2[\overline{\tilde{b}(1/\overline{z})} + k\tilde{b}(1/\overline{z})]^{-2}$$
$$= -4(1-k^2)\left\{\frac{1-z}{\sqrt{z}}\left[\left(\frac{\sqrt{z}+1}{\sqrt{z}-1}\right)^{\sigma} - \left(\frac{\sqrt{z}-1}{\sqrt{z}+1}\right)^{\sigma}\right] + k\,\overline{\frac{1-z}{\sqrt{z}}\left[\left(\frac{\sqrt{z}+1}{\sqrt{z}-1}\right)^{\sigma} - \left(\frac{\sqrt{z}-1}{\sqrt{z}+1}\right)^{\sigma}\right]}\right\}^{-2}$$

Comparing the two representations for f on $|z| = 1$, we have

$$1/f(e^{i\theta}) = -(\sin\tfrac{1}{2}\theta)^2[e^{\frac{1}{2}i\sigma\pi}(\cot\tfrac{1}{4}\theta)^{\sigma} + e^{-\frac{1}{2}i\sigma\pi}(\tan\tfrac{1}{4}\theta)^{\sigma}]^2$$
$$= \frac{-(\sin\tfrac{1}{2}\theta)^2}{1-k^2}[(-ie^{-\frac{1}{2}i\sigma\pi} + ike^{\frac{1}{2}i\sigma\pi})(\cot\tfrac{1}{4}\theta)^{\sigma} + (ie^{\frac{1}{2}i\sigma\pi} - ike^{-\frac{1}{2}i\sigma\pi})(\tan\tfrac{1}{4}\theta)^{\sigma}]^2 .$$

This is possible only if $\pm\sqrt{1-k^2}\,e^{\frac{1}{2}i\sigma\pi} = -ie^{-\frac{1}{2}i\sigma\pi} + ike^{\frac{1}{2}i\sigma\pi}$. Therefore $\sigma = \frac{1}{\pi}\arccos k = \varkappa$ and $a_2 = 2 - 4\varkappa^2$.

<u>REMARK</u>. The construction of the expressions c and d in the proof of Theorem 14.13 is a general procedure that has been used by R. Kühnau [K10,K12] to attack a number of problems.

<u>EXERCISE</u>. Adapt Theorem 14.11 to the class $S_{K,R}$.

<u>PROBLEM</u>. Solve the problem $\max\limits_{S_{K,R}} |a_2|$.

The family S_K is an example of a <u>linearly</u> <u>invariant</u> family. That is, if $f \in S_K$, then F_ζ defined by

$$F_\zeta(z) = \frac{f([z+\zeta]/[1+\overline{\zeta}z]) - f(\zeta)}{(1-|\zeta|^2)f'(\zeta)}$$

belongs to S_K for $\zeta \in U$. Application of Theorem 14.13 to the second coefficient of F_ζ yields the following consequence:

COROLLARY 14.14. If $f \in S_K$, then

$$\left|(1-|\zeta|^2)\frac{\zeta f''(\zeta)}{f'(\zeta)} - 2|\zeta|^2\right| < 4 - 8\pi^{-2}(\arccos k)^2 \quad \text{for } \zeta \in U .$$

We may view the inequality of Corollary 14.14 as a necessary condition for $f \in S$ to have a K-q.c. extension to $\bar{C}$. In Chapter 15 we shall consider a similar condition that is sufficient.

We turn now to a nonlinear problem. Recall the definitions of L^2 and $|L|^2$ from Chapter 11. We continue to identify functions in $\Sigma_K \cup \Sigma_{K,r}$ and $S_\Sigma \cup S_{K,R}$ with their restrictions in Σ and S, respectively.

THEOREM 14.15 (Kühnau [K9], McLeavey [M2]). If $g \in \Sigma_{K,r}$ and $L \in H'(|z|>1)$, then

$$\left|L^2\left(\log \frac{g(z)-g(\zeta)}{z-\zeta}\right)\right| \le |L|^2\left(\sum_{n=1}^{\infty} \frac{k+r^{2n}}{1+kr^{2n}} \frac{(z\bar{\zeta})^{-n}}{n}\right) .$$

In particular, if $g \in \Sigma_K$ and $L \in H'(|z|>1)$, then

$$\left|L^2\left(\log \frac{g(z)-g(\zeta)}{z-\zeta}\right)\right| \le k\, |L|^2\left(\log \frac{1}{1-1/(z\bar{\zeta})}\right) .$$

Similarly, if $f \in S_{K,R}$ and $L \in H'(U)$, then

$$\left|L^2\left(\log \frac{f(z)-f(\zeta)}{z-\zeta} \frac{z\zeta}{f(z)f(\zeta)}\right)\right| \le |L|^2\left(\sum_{n=1}^{\infty} \frac{1+kR^{2n}}{k+R^{2n}} \frac{(z\bar{\zeta})^{n}}{n}\right),$$

and if $f \in S_K$ and $L \in H'(U)$, then

$$\left|L^2\left(\log \frac{f(z)-f(\zeta)}{z-\zeta} \frac{z\zeta}{f(z)f(\zeta)}\right)\right| \le k\, |L|^2\left(\log \frac{1}{1-z\bar{\zeta}}\right) .$$

Proof. For the first inequality we may assume, as in the proof of Theorem 11.3, that $L \neq 0$ and that $g \in \Sigma'_{K,r}$ is an extremal function for the problem

$$\max_{\Sigma'_{K,r}} \operatorname{Re} L^2\left(\log \frac{g(z)-g(\zeta)}{z-\zeta}\right) = \max_{\Sigma_{K,r}} \operatorname{Re} L^2\left(\log \frac{g(z)-g(\zeta)}{z-\zeta}\right).$$

As before, the complex Gâteaux derivative of the functional at g is defined by

$$L(h;g) = L^2\left(\frac{h(z)-h(\zeta)}{g(z)-g(\zeta)}\right).$$

Therefore

$$L(\Psi(\omega,g);g) = L(1/(g-\omega);g) = -L^2(1/([g(z)-\omega][g(\zeta)-\omega])) = -[L(1/(g-\omega))]^2$$

and $$J(w) = \int^w \sqrt{L(\Psi(\omega,g);g)}\,d\omega = i\int^w L(1/(g-\omega))\,d\omega .$$

By Theorem 4.1 we may represent $L(h) = \int_{|z|=\rho} h(z)\ell(z)\,dz$ for some function $\ell(z)$ analytic and not identically zero in $|z| \le \rho$, $\rho > 1$. Therefore

$$J(w) = i\int^w \int_{|z|=\rho} \frac{\ell(z)}{g(z)-\omega}\,dz\,d\omega = -i\int_{|z|=\rho} (\log[g(z)-w])\,\ell(z)\,dz$$

is an analytic function of w inside $g(|z|=\rho)$. By Corollary 13.4,

$$b = J\circ g - k\overline{J\circ g}$$

is analytic in $r < |z| < 1$ except possibly at the zeros of $L(1/[g - g(\cdot)])$. These points are clearly removable singularities; so we may expand

$$b(z) = \sum_{n=-\infty}^{\infty} \beta_n z^n \quad \text{in} \quad r < |z| < 1 .$$

We now apply conformal boundary variations, as in the proof of Theorem 14.8. Then it follows from Schiffer's fundamental lemma (Theorem C.4) that $\mathbb{C}-g(|z| > r)$ consists of finitely many analytic arcs satisfying the differential equation

$$dJ(w) = iL(1/(g-w))\,dw = \text{real}$$

(cf. Theorem 11.3). That is,

$$db = d(J\circ g - k\,\overline{J\circ g}) = \text{real}$$

on $|z| = r$, except possibly at finitely many singularities. Consequently, $b(z) - b(r)$ is real and continuous on $|z| = r$, except possibly for finitely many singularities, at which it is bounded. By the Schwarz reflection principle, these singularities are removable and b extends analytically to $r^2 < |z| < 1$ through the identity

$$b(z) = \overline{b(r^2/\bar{z})} + \text{imaginary constant}.$$

It follows from this identity that

$$\beta_{-n} = r^{2n}\overline{\beta_n} \qquad \text{for all } n \neq 0.$$

Moreover, since $(1-k^2)J\circ g = b + k\bar{b}$ on $|z| = 1$, the Fourier coefficients

$$\int_{|\zeta|=1} J\circ g(\zeta)\zeta^{n-1}d\zeta = \frac{k+r^{2n}}{1+kr^{2n}}\overline{\int_{|\zeta|=1} J\circ g(\zeta)\zeta^{-n-1}d\zeta} \qquad \text{for all } n \neq 0.$$

On the other hand,

$$\int_{|\zeta|=1} J\circ g(\zeta)\zeta^{n-1}d\zeta = -i\int_{|\zeta|=1}\int_{|z|=\rho}\left[\log\frac{g(z)-g(\zeta)}{z-\zeta} + \log(z-\zeta)\right]\ell(z)\zeta^{n-1}dz\,d\zeta$$

$$= \begin{cases} \dfrac{-i}{n}L(z^n) & \text{for } n < 0 \\ -i\displaystyle\int_{|\zeta|=1} L\left(\log\frac{g(z)-g(\zeta)}{z-\zeta}\right)\zeta^{n-1}d\zeta & \text{for } n > 0. \end{cases}$$

Thus $-i\displaystyle\int_{|\zeta|=1} L\left(\log\frac{g(z)-g(\zeta)}{z-\zeta}\right)\zeta^{n-1}d\zeta = \frac{k+r^{2n}}{1+kr^{2n}}\overline{\frac{-i}{-n}L(z^{-n})}$ for all $n > 0$.

Therefore each extremal function g satisfies the identity

$$L\left(\log\frac{g(z)-g(\zeta)}{z-\zeta}\right) = \sum_{n=1}^{\infty}\frac{k+r^{2n}}{1+kr^{2n}}\frac{1}{n}\overline{L(z^{-n})}\,\zeta^{-n} \quad \text{for } |\zeta| > 1.$$

The extreme value of the functional is obtained by applying L to both sides, and the first inequality of the theorem is a consequence.

The second inequality of the theorem follows from the first by letting $r \to 0$, since $\Sigma_K\big|_{|z|>r} \subset \Sigma_{K,r}$ for every r. The remaining inequalities of the theorem follow from applying the former ones to $g(z) = 1/f(1/z)$.

EXERCISE. Imitate the pertinent parts of the above proof to obtain the second inequality of Theorem 14.15 directly.

We shall note a number of consequences of Theorem 14.15:

COROLLARY 14.16. If $g \in \Sigma_{K,r}$, $f \in S_{K,R}$, and $p \in C$, then

$$|L^2\Big(\Big[\frac{g(z)-g(\zeta)}{z-\zeta}\Big]^p\Big)| \le |L|^2\Big(\Big[1-\frac{1}{z\bar\zeta}\Big]^{-k|p|}\prod_{n=1}^{\infty}\Big[1-\frac{r^{2n}}{z\bar\zeta}\Big]^{(k^2-1)(-k)^{n-1}|p|}\Big)$$

for all $L \in H'(|z|>1)$, and

$$|L^2\Big(\Big[\frac{f(z)-f(\zeta)}{z-\zeta}\,\frac{z\zeta}{f(z)f(\zeta)}\Big]^p\Big)| \le |L|^2\Big([1-z\bar\zeta]^{-k|p|}\prod_{n=1}^{\infty}\Big[1-\frac{z\bar\zeta}{R^{2n}}\Big]^{(k^2-1)(-k)^{n-1}|p|}\Big),$$

for all $L \in H'(U)$. For $g \in \Sigma_K$ and $f \in S_K$ the infinite products are omitted.

Proof. Apply Theorem 11.16 with $\varphi(z) = e^{pz}$. Cf. Corollary 11.17.

EXERCISE. Interpret the inequalities of Theorem 14.15 and Corollary 14.16 for the Goluzin functional defined by $L(h) = \sum_{m=1}^{N}\lambda_m h(z_m)$. Compare to Corollary 11.5.

COROLLARY 14.17. If $g \in \Sigma_{K,r}$ and $f \in S_{K,R}$, then

$$|\log g'(z)| \le -k\log\Big(1-\frac{1}{|z|^2}\Big) - (1-k^2)\log\prod_{n=1}^{\infty}\Big(1-\frac{r^{2n}}{|z|^2}\Big)^{(-k)^{n-1}} \quad \text{for } |z|>1,$$

$$\Big|\log\frac{z^2 f'(z)}{f(z)^2}\Big| \le -k\log(1-|z|^2) - (1-k^2)\log\prod_{n=1}^{\infty}\Big(1-\frac{|z|^2}{R^{2n}}\Big)^{(-k)^{n-1}} \quad \text{for } z \in U,$$

$$|\{g;z\}| \le \frac{6k}{(|z|^2-1)^2} + 6(1-k^2)\sum_{n=1}^{\infty} \frac{(-k)^{n-1} r^{2n}}{(|z|^2 - r^{2n})^2} \qquad \text{for } |z|>1 ,$$

$$|\{f;z\}| \le \frac{6k}{(1-|z|^2)^2} + 6(1-k^2)\sum_{n=1}^{\infty} \frac{(-k)^{n-1} R^{2n}}{(R^{2n} - |z|^2)^2} \qquad \text{for } z \in U .$$

For $g \in \Sigma_K$ and $f \in S_K$ the terms with infinite products or infinite series do not appear.

<u>Proof</u>. Apply Theorem 14.15 with $L(h) = h(z)$ and $L(h) = \sqrt{6}h'(z)$. Cf. Corollaries 11.5 and 11.6.

<u>COROLLARY 14.18</u>. Let $g \in \Sigma_{K,r}$ have Grunsky matrix $[\gamma_{mn}]$ generated by

$$\log \frac{g(z)-g(\zeta)}{z-\zeta} = \sum_{m,n=1}^{\infty} \gamma_{mn} z^{-m} \zeta^{-n} \qquad |z|>1 , \quad |\zeta|>1 ,$$

and $f \in S_{K,R}$ have Grunsky matrix $[c_{mn}]$ generated by

$$\log \frac{f(z)-f(\zeta)}{z-\zeta} = \sum_{m,n=0}^{\infty} c_{mn} z^m \zeta^n \qquad z,\zeta \in U .$$

Then for all complex sequences $\{\lambda_n\}$ with $\limsup_{n\to\infty} |\lambda_n|^{1/n} < 1$,

$$\Big| \sum_{m,n=1}^{\infty} \gamma_{mn}\lambda_m\lambda_n \Big| \le \sum_{n=1}^{\infty} \frac{k+r^{2n}}{1+kr^{2n}} \frac{|\lambda_n|^2}{n} ,$$

$$\Big| \sum_{m,n=1}^{\infty} c_{mn}\lambda_m\lambda_n \Big| \le \sum_{n=1}^{\infty} \frac{1+kR^{2n}}{k+R^{2n}} \frac{|\lambda_n|^2}{n} ,$$

and

$$\sum_{m=1}^{N} m \frac{1+kr^{2m}}{k+r^{2m}} \Big| \sum_{n=1}^{N} \gamma_{mn}\lambda_n \Big|^2 \le \sum_{n=1}^{N} \frac{k+r^{2n}}{1+kr^{2n}} \frac{|\lambda_n|^2}{n} ,$$

$$\sum_{m=1}^{N} m \frac{k+R^{2m}}{1+kR^{2m}} \Big| \sum_{n=1}^{N} c_{mn}\lambda_n \Big|^2 \le \sum_{n=1}^{N} \frac{1+kR^{2n}}{k+R^{2n}} \frac{|\lambda_n|^2}{n} .$$

For $g \in \Sigma_K$ and $f \in S_K$, let $r = 0$ and $R \to \infty$.

<u>Proof</u>. The proof is similar to the proofs of Corollaries 11.7 and 11.9.

<u>EXAMPLE</u>. Let $g \in \Sigma_{K,r}$ and $g(z) = z + \sum_{n=0}^{\infty} b_n z^{-n}$ for $|z| > 1$.

As in the example on p. 120, by choosing $\lambda_1 = 1$, $\lambda_2 = \dots = \lambda_N = 0$ in Corollary 14.18 and letting $N \to \infty$, one obtains the area inequality

$$\sum_{m=1}^{\infty} m|b_m|^2 \le \left(\frac{k+r^2}{1+kr^2}\right)^2 .$$

That is, the area of $\mathbb{C}-g(|z|>1)$ is at least (cf. Theorem B.1)

$$\pi(1-k^2)(1-r^4)/(1+kr^2)^2 .$$

For $g \in \Sigma_K$ set $r = 0$.

EXERCISE. Deduce the bounds of Theorems 14.5 and 14.8 from the above example.

REMARK. Based on the Grunsky inequalities of Corollary 14.18 for S_K and the above area inequality for Σ_K, R. Kühnau [K11] has applied the argument of Charzynski and Schiffer for Theorem 11.11 to obtain the following estimate, which is not sharp:

If $f \in S_K$, $f(\infty) = \infty$, and $f(z) = z + \sum_{n=2}^{\infty} a_n z^n$ for $z \in U$, then

$$|a_4| \le \begin{cases} \frac{2}{3}k + \frac{4}{\sqrt{3}}k^2 + \frac{10}{3}k^3 & \text{for } 0 < k < \sqrt{7/15} \\ \frac{2}{3}k + \frac{10}{3}k^3 & \text{for } \sqrt{7/15} \le k < 1 . \end{cases}$$

One obtains the same estimate with k replaced by $\frac{1+kR}{k+R}$ throughout, if $f \in S_{K,R}$ and $f \ne \infty$.

THEOREM 14.19. Let $g \in \Sigma'_{K,r}$, $g \ne 0$, and $g(z) = z + \sum_{n=1}^{\infty} b_n z^{-n}$ for $|z| > 1$. Then

$$|b_2| \le \tfrac{2}{3}(k+r^3)/(1+kr^3) .$$

Furthermore, if $g \in \Sigma'_K$ and $g(0) = 0$, then

$$|b_2| \le \tfrac{2}{3}k .$$

Proof. Apply the Grunsky inequalities of Corollary 14.18, with $\lambda_3 = 1$ and $\lambda_n = 0$ for $n \neq 3$, to $\sqrt{g(z^2)} \in \Sigma'_{K,\sqrt{r}}$. Then

$$\frac{1}{2}|b_2| = |\gamma_{33}| \leq \frac{1}{3}(k + r^3)/(1 + kr^3) .$$

If $g \in \Sigma'_K$ and $g(0) = 0$, one may let $r \to 0$.

The above proof does not extend immediately to the unrestricted families $\Sigma_{K,r}$ and Σ_K. However, by another method O. Lehto [L4] has shown that $|b_2| \leq \frac{2}{3}k$ for $g \in \Sigma_K$, in general.

CHAPTER 15. Sufficient conditions for q.c. extensions

The estimates for various functionals in Chapter 14 can be viewed as necessary conditions for conformal mappings to have q.c. extensions. Necessary and sufficient conditions are known in terms of boundary behavior. In particular, if $f \in S$ maps U onto a Jordan domain whose boundary curve C has the property that

$$|z_1 - z_2||z_3 - z_4| \le B|z_1 - z_3||z_2 - z_4|$$

uniformly for some constant B, whenever $z_1, z_2, z_3, z_4 \in C$ in that order, then f has a q.c. extension to $\bar{C}$ (Ahlfors [A4]). Roughly speaking, this geometric condition permits C to have corners, but excludes cusps. However, the smallest K, as a function of B, for which a K-q.c. extension exists is not known.

The conditions of Theorem 14.15 carry a considerable amount of necessary information for existence of a K-q.c. extension. It follows from the work of M. Schiffer [S9], G. Springer [S18], and Chr. Pommerenke [P8] that they are also sufficient for a q.c. extension to exist. It is an intriguing problem whether the conditions of Theorem 14.15 are sufficient for a K-q.c. extension to exist with the corresponding K (cf. Corollary 11.8).

PROBLEM. Are the conditions of Theorem 14.15 sufficient for a K-q.c. extension to exist with the same K ?

Some quantitative information is known. We refer to the inequality for zf''/f' in Corollary 14.14 and to the inequality

$$(1-|z|^2)^2|\{f;z\}| \le 6k$$

in Corollary 14.17 for $f \in S_K$, $z \in U$. There are corresponding sufficient conditions of a similar form for a K-q.c. extension to exist:

<u>THEOREM 15.1</u> (Ahlfors [A7]). Let $f \in S$ and assume

$$\text{(A)} \qquad \left|(1-|z|^2)\frac{zf''(z)}{f'(z)} - c|z|^2\right| \le k, \qquad z \in U,$$

or

$$\text{(B)} \qquad |(1-|z|^2)^2\{f;z\} - 2c(1-c)\bar{z}^2| \le 2k|1-c|, \quad z \in U,$$

for some constant $c \in C$ with $|c| \le k < 1$. Then there exists an $F \in S_K$, $K = (1+k)/(1-k)$, such that $F|_U = f$. In case (A) an F exists with $F(\infty) = \infty$.

<u>Proof</u>. Assume first that f is analytic and univalent in a domain containing $\bar{U}$. Represent F in the form

$$F(z) = \begin{cases} f(z) & \text{for } |z| \le 1 \\ \overline{g(1/\bar{z})} & \text{for } |z| > 1 \end{cases}$$

where g is to be K-q.c. in U and $g = \bar{f}$ on $|z| = 1$. Furthermore, set

$$g = \overline{f + \tau f'}$$

where $\tau = 0$ on $|z| = 1$. The K-qcty condition is then

$$|(1+\tau_z)f' + \tau f''| = |g_{\bar{z}}| \le k|g_z| = k|\tau_{\bar{z}} f'|$$

or

$$|1 + \tau_z + \tau f''/f'| \le k|\tau_{\bar{z}}|.$$

If we force g to be locally homeomorphic in U, then F will be a locally homeomorphic mapping of $\bar{C}$ into $\bar{C}$. The only such mappings are then global homeomorphisms of $\bar{C}$ onto $\bar{C}$. To this end, it is sufficient that τ be smooth and $\tau_{\bar{z}} \ne 0$ at points where τ is finite, and $1/g$ be smooth and $(1/\tau)_{\bar{z}} \ne 0$ at the point(s) where τ is infinite.

In case (A), let

$$\tau(z) = \frac{1-|z|^2}{(1-c)\bar{z}} .$$

Then the K-qcty condition is precisely assumption (A), τ vanishes on $|z|=1$, $\tau_{\bar{z}} = -1/[(1-c)\bar{z}^2] \neq 0$ for $z \neq 0$, and $(1/\tau)_{\bar{z}} = 1-c \neq 0$ at the origin. Since $\tau(0) = \infty$, it is evident that $F(\infty) = \infty$.

In case (B), let

$$\tau(z) = \frac{1-|z|^2}{(1-c)\bar{z} - \frac{1}{2}(1-|z|^2)f''(z)/f'(z)} .$$

Then the K-qcty condition is precisely assumption (B), τ vanishes only on $|z| = 1$, $\tau_{\bar{z}} = (c-1)\tau^2/(1-|z|^2)^2 \neq 0$ at the points where $\tau \neq \infty$, and $(1/\tau)_{\bar{z}} = (1-c)/(1-|z|^2)^2 \neq 0$ at the point(s) where $\tau = \infty$. Since F is now a global homeomorphism, it is a consequence that F (and τ) are infinite at precisely one point.

The proof has been completed under regularity assumptions on f at $|z| = 1$. For arbitrary f we shall apply what has been proved to the functions $f_r(z) = f(rz)/r$, $0<r<1$. By letting $r \to 1$, the theorem will then follow, in general, from the compactness of S_K. It remains only to show that if f satisfies (A) or (B), then so does f_r.

If f satisfies (A), then

$$\left| \frac{zf_r''(z)}{f_r'(z)} - \frac{cr^2|z|^2}{1-r^2|z|^2} \right| \leq \frac{k}{1-r^2|z|^2} .$$

By the triangle inequality, f_r also satisfies (A) since $|c| \leq k$ and

$$\frac{|z|^2}{1-|z|^2} - \frac{r^2|z|^2}{1-r^2|z|^2} = \frac{1}{1-|z|^2} - \frac{1}{1-r^2|z|^2} .$$

Similarly, if f satisfies (B), then

$$\left| \{f_r;z\} - \frac{2c(1-c)r^4\bar{z}^2}{(1-r^2|z|^2)^2} \right| \le \frac{2k|1-c|r^2}{(1-r^2|z|^2)^2} \, .$$

By the trianglé inequality, f_r also satisfies (B) since $|c| \le k$ and

$$\left| \frac{\bar{z}^2}{(1-|z|^2)^2} - \frac{r^4\bar{z}^2}{(1-r^2|z|^2)^2} \right| \le \frac{1}{(1-|z|^2)^2} - \frac{r^2}{(1-r^2|z|^2)^2} \, .$$

APPENDIX A. Some convexity theory

Suppose A is a subset of a linear (vector) space over $\mathbb{C}$ or $\mathbb{R}$. A is <u>convex</u> if $[\lambda x + (1-\lambda)y] \in A$ for all $x, y \in A$ and all $\lambda \in (0,1)$. A point $\xi \in A$ is an <u>extreme point</u> of A if $\xi \neq \lambda x + (1-\lambda)y$ for all distinct $x, y \in A$ and all $\lambda \in (0,1)$. We denote the set of extreme points of A by E_A.

If A is a subset of a linear topological space, then the <u>closed convex hull</u> of A, denoted by $\overline{co}\,A$, is the smallest (intersection) of all closed convex sets containing A.

<u>THEOREM A.1</u> (Krein - Milman theorem). If A is a compact subset of a locally convex linear topological space, then $\overline{co}\,A = \overline{co}\,E_A$. If, in addition, $\overline{co}\,A$ is compact, then $E_{\overline{co}\,A} \subset A$.

In particular, we note that $E_A \neq \emptyset$ if A is nonempty and compact. A proof of the Krein - Milman theorem can be found in the books by N. Dunford and J. T. Schwartz [D1] and N. Bourbaki [B6].

Closed convex sets and their extreme points are preserved by linear homeomorphisms:

<u>THEOREM A.2</u>. Let X and Y be linear topological spaces and $\mathcal{L}: X \to Y$ a linear homeomorphism. If $A \subset X$, then $\overline{co}\,\mathcal{L}(A) = \mathcal{L}(\overline{co}\,A)$ and $E_{\overline{co}\,\mathcal{L}(A)} = \mathcal{L}(E_{\overline{co}\,A})$.

The proof is an elementary exercise that we recommend to the reader.

For solving linear extremal problems over compact sets, it is sufficient to test only the extreme points:

THEOREM A.3. If A is a compact subset of a locally convex linear topological space X and x' is a continuous linear functional on X, then $\max_A \Re x' = \max_{E_A} \Re x'$. If, in addition, $\overline{\text{co}}\, A$ is compact, then $\max_A \Re x' = \max_{\overline{\text{co}}\, A} \Re x' = \max_{E_{\overline{\text{co}}\, A}} \Re x'$.

Proof. Assume $A \neq \emptyset$. Then $B = \{x \in A: \Re x'(x) = \max_A \Re x'\}$ is nonempty and compact, so $E_B \neq \emptyset$. Let $\xi \in E_B$. If $\xi = \lambda x + (1-\lambda)y$ for some distinct $x, y \in A$ and $\lambda \in (0,1)$, then $x, y \in B$ since $\Re x'(\xi)$ is a maximum. Consequently, $\xi \in E_A$ and $\max_A \Re x'$ is attained in E_A.

If $\overline{\text{co}}\, A$ is compact, then the above argument shows that $\max_{\overline{\text{co}}\, A} \Re x'$ is attained in $E_{\overline{\text{co}}\, A} \subset A$.

Observe that Theorem A.3 remains true with any continuous convex function in place of $\Re x'$.

By replacing x' by $-x'$, ix', $-ix'$, we note that the problem $\max_A \Re x'$ includes the problems $\min_A \Re x'$, $\min_A \Im x'$, $\max_A \Im x'$. Moreover, if $\max_A |x'| = |x'(\xi)|$ and $x_1' = \frac{|x'(\xi)|}{x'(\xi)} x'$, then $\max_A \Re x_1' = |x'(\xi)| = \max_A |x'|$, so that the problem $\max_A |x'|$ is also included.

The Krein-Milman theorem says that compact convex sets can be reconstructed from just their extreme points. Choquet's theorem says that the construction can be carried out by integrating the extreme points with respect to probability measures.

THEOREM A.4 (Choquet's theorem). Suppose that A is a compact convex subset of a metrizable, locally convex, real linear space X and $x \in A$. Then there exists a probability measure μ_x supported by E_A such that

$$x = \int_{E_A} y \, d\mu_x(y) .$$

Moreover, if A is a simplex, then μ_x is unique.

By a <u>probability</u> <u>measure</u> we mean a nonnegative regular Borel measure with total mass one. Under the metrizability hypothesis of Choquet's theorem the set E_A is a G_δ set. For a proof of Choquet's theorem see the books by E. M. Alfsen [A9] and R. R. Phelps [P5] and the references there. Actually, the existence of a probability measure μ_x supported in the closure of E_A is a consequence of the Krein - Milman theorem, essentially the weak compactness of the set of probability measures. The uniqueness of μ_x lies a little deeper. We must yet define the notion of simplex in the above context.

In the definition of simplex we assume that A lies in a hyperplane that misses the origin. Then $B = \{\alpha x: \alpha \geq 0 , x \in A\}$ is a cone generated by A and induces a partial ordering on $D = \{x - y: x , y \in B\}$ defined by $x \geq y$ if $x - y \in B$. A is called a (Choquet) <u>simplex</u> if this partial ordering of D is a lattice, i.e., if each pair $x , y \in D$ has a least upper bound $x \vee y \in D$. z is a least upper bound of x and y if $z \geq x$, $z \geq y$, and if $z \leq w$ whenever $w \geq x$, $w \geq y$.

As an example, X might be the space of all harmonic functions on U with the topology of locally uniform convergence, and $A = \{u \in X: u > 0 , u(0) = 1\}$. Then X is a metrizable locally convex real linear space, A is a compact convex subset, and the cone $B = \{u \in X: u > 0\}$ induces a lattice structure on $D = \{u - v: u, v > 0;\ u, v \in X\}$. Consequently, A is a simplex.

The existence of the measure μ in Theorem 1.6′ follows immediately from Choquet's theorem. However, the uniqueness uses also the fact that $u_1, u_2 \in D$ have a least harmonic majorant in D.

APPENDIX B. Coefficient and distortion theorems

Let $$S = \{f \in H_u(U) : f(z) = z + \sum_{n=2}^{\infty} a_n z^n\}$$

and $$\Sigma = \{g \in H_u(|z|>1) : g(z) = z + \sum_{n=0}^{\infty} b_n z^{-n}\} .$$

In this appendix we shall collect some standard information about S and Σ.

THEOREM B.1 (area theorem; Gronwall [G9]). If $g(z) = z + \sum_{n=0}^{\infty} b_n z^{-n} \in \Sigma$, then

$$\sum_{n=1}^{\infty} n|b_n|^2 \le 1 .$$

In particular, $|b_1| \le 1$ with equality iff $g(z) = z + b_0 + e^{i\alpha}/z$.

Proof. If the circle $|z| = r > 1$ is oriented positively with respect to the origin, then by Green's theorem

$$0 < \frac{1}{\pi} \iint_{\mathrm{Int}\, g(|z|=r)} du\, dv = \frac{1}{2\pi} \int_{g(|z|=r)} u\, dv - v\, du = \Re e\, \frac{1}{2\pi i} \int_{|z|=r} \bar{g}\, dg$$

$$= \Re e\, \frac{1}{2\pi i} \int_{|z|=r} \left[\frac{r^2}{z} + \sum_{m=0}^{\infty} \overline{b_m}\left(\frac{z}{r^2}\right)^m\right]\left[z - \sum_{n=1}^{\infty} n b_n z^{-n}\right]\frac{dz}{z} = r^2 - \sum_{n=1}^{\infty} n|b_n|^2 r^{-2n} .$$

Let $r \to 1$.

THEOREM B.2 (Bieberbach [B2]). If $f(z) = z + \sum_{n=2}^{\infty} a_n z^n \in S$, then $|a_2| \le 2$. Equality occurs iff $f(z) = z/(1-\eta z)^2$, $|\eta| = 1$.

Proof. Apply the estimate for the first coefficient from Theorem B.1 to the odd function $g(z) = z[z^2 f(1/z^2)]^{-\frac{1}{2}} =$ $= z - \frac{1}{2}a_2 z^{-1} + \dots$, which is in Σ.

In 1907, P. Koebe [K2] proved the existence of a positive constant $\varkappa$ such that $\bigcap_{f \in S} f(U)$ contains the disk $|w| < \varkappa$. In 1916, L. Bieberbach [B2] obtained $\frac{1}{4}$ as the best possible value for $\varkappa$.

<u>THEOREM B.3</u> ($\frac{1}{4}$-theorem). $\bigcap_{f \in S} f(U) = \{|w| < \frac{1}{4}\}$.

<u>Proof</u>. If $f \in S$ and $w \notin f(U)$, then $f(z)/[1 - \frac{1}{w} f(z)] = z + (a_2 + \frac{1}{w}) z^2 + \dots$ also belongs to S. By Theorem B.2, we have $|a_2 + \frac{1}{w}| \le 2$ and $\frac{1}{|w|} \le 2 + |a_2| \le 4$. Consequently, $\bigcap_{f \in S} f(U) \supset \{|w| < \frac{1}{4}\}$. The Koebe functions $k(z) = z/(1 - \eta z)^2$, $|\eta| = 1$, belong to S, and $k(U)$ omits the line $\{t\bar{\eta}: -\infty < t \le -\frac{1}{4}\}$. Therefore $\bigcap_{f \in S} f(U) \subset \{|w| < \frac{1}{4}\}$.

Since the estimate $|a_2| \le 2$ from Theorem B.2 was used in the above proof, the Koebe functions are the only ones whose image contains a boundary point on the circle $|w| = \frac{1}{4}$.

There are related results for functions which need not be univalent:

> If $f(z) = z + \sum_{n=2}^{\infty} a_n z^n$ is analytic in $\bar{U}$, then there exists an absolute constant $L > 0$ such that $f(U)$ contains some disk of radius L.

E. Landau [L2] proved that the best possible value of L is at least $1/16$. It is now known to be between $\frac{1}{2}$ and $.56\ldots$. A. Bloch [B5] proved the following stronger theorem:

> If $f(z) = z + \sum_{n=2}^{\infty} a_n z^n$ is analytic in $\bar{U}$, then there exists an absolute constant $B > 0$ such that $f(U)$ contains some disk of radius B which

is the one-to-one image of some subdomain of U . It is known that the best value of B is between .43... and .47... . L. V. Ahlfors and H. Grunsky [A8] believe that it is the latter.

THEOREM B.4 (Bieberbach distortion theorem [B2]). If $f \in S$ and $|z| = r < 1$, then

(a) $r/(1+r)^2 \le |f(z)| \le r/(1-r)^2$

(b) $(1-r)/(1+r)^3 \le |f'(z)| \le (1+r)/(1-r)^3$.

Proof. Let $f \in S$ and $k(z) = z/(1+e^{-i\theta}z)^2$. Since $b = 4r/(1+r)^2 < 1$ for $0 < r < 1$, Theorem B.3 implies that $\frac{1}{4} \le |\frac{1}{b} f \circ k^{-1} \circ (bk)(e^{i\theta})| = \frac{1}{b}|f(re^{i\theta})|$, giving the lower bound in (a) . For the upper bound in (a), fix $|\zeta| < 1$ and define

$$F_\zeta(z) = \frac{f([z+\zeta]/[1+\bar{\zeta}z]) - f(\zeta)}{(1-|\zeta|^2)f'(\zeta)}$$

Then $F_\zeta \in S$ and $|F_\zeta(-\zeta)| \ge r/(1+r)^2$ for $|\zeta| = r$, by what is already proved. Therefore $|\frac{\partial}{\partial r}\log f(re^{i\theta})| \le \frac{\partial}{\partial r}\log[r/(1-r)^2]$, and by integration $|f(re^{i\theta})| \le [r/(1-r)^2]\lim_{r_o\to 0}(1-r_o)^2|f(r_o e^{i\theta})/r_o|$. Both bounds in (b) follow from applying (a) to $F_\zeta(-\zeta)$.

REMARKS. Again the Koebe functions are the only ones for which (a) and (b) are sharp, since the estimates were based on the $\frac{1}{4}$-theorem.

As a consequence of (a), the family S is locally uniformly bounded, hence normal, and compactness follows easily.

The Bieberbach conjecture asserts that $|a_n| \le n$ for all

$f(z) = z + \sum_{n=2}^{\infty} a_n z^n$ in S. It has been verified for $n = 2$ (above, Bieberbach [B2]), for $n = 3$ (Löwner [L6]), for $n = 4$ (Garabedian and Schiffer [G3]), for $n = 5$ (Pederson and Schiffer [P3]), and for $n = 6$ (Pederson [P2], Ozawa [O1]).

Some additional information is known. Except for the Koebe functions, $\limsup_{n\to\infty} |a_n|/n < 1$ (Hayman [H2]). Therefore $|a_n| \le n$ for all n sufficiently large, depending on f. If $|a_2|$ is sufficiently close to 2, depending on n, then $|a_n| \le n$ (Garabedian, Ross, and Schiffer [G1, G4]). On the other hand, if $|a_2| \le 1.05$, then $|a_n| \le n$ (Aharonov [A2], Il'ina [I1]). Currently, the best uniform estimate is $|a_n| \le \sqrt{7/6}\, n \approx 1.08\, n$ (FitzGerald [F1]).

A distortion theorem results from each successive coefficient for which the Bieberbach conjecture is verified:

<u>THEOREM B.5</u> (Landau [L1]). If the Bieberbach conjecture $\max_S |a_\nu| = \nu$ is true for $2 \le \nu \le n$, then

$$|f^{(n)}(z)| \le (n!)(n + |z|)/(1-|z|)^{n+2} \qquad \text{for } f \in S \text{ and } z \in U .$$

<u>Proof</u>. Let $f \in S$, $0 \le r < 1$, and $F(z) = f\left(\frac{z + r}{1 + rz}\right) = \sum_{\nu=0}^{\infty} c_\nu z^\nu$. Then $f(z) = F\left(\frac{z - r}{1 - rz}\right) = \sum_{\nu=0}^{\infty} c_\nu \left(\frac{z - r}{1 - rz}\right)^\nu$ and

$$\frac{f^{(n)}(r)}{n!} = \sum_{\nu=1}^{n} \frac{c_\nu}{(n-\nu)!} \frac{d^{n-\nu}}{dz^{n-\nu}} (1 - rz)^{-\nu}\Big|_{z=r} = \sum_{\nu=1}^{n} c_\nu P_\nu(r,n)$$

where $P_\nu(r,n)$ is nonnegative and independent of f. Since $[F(z) - F(0)]/F'(0) \in S$, the coefficients

$$|c_\nu| \le \nu|F'(0)| = \nu(1-r^2)|f\ (r)| \le \nu[(1+r)/(1-r)]^2$$

for $2 \le \nu \le n$ by the hypothesis and Theorem B.4(b). Therefore

$$\frac{|f^{(n)}(r)|}{n!} \le \sum_{\nu=1}^{n} \nu[(1+r)/(1-r)]^2 P_\nu(r,n) .$$

To identify the last sum, we consider the special case of $k(z) = z/(1-z)^2$. The coefficients c_ν of $k\left(\frac{z+r}{1+rz}\right) = r/(1-r)^2 + [(1+r)/(1-r)]^2 z/(1-z)^2$ are just $\nu[(1+r)/(1-r)]^2$ for $\nu \ge 1$. Therefore

$$\begin{aligned} \frac{|f^{(n)}(r)|}{n!} &\le \sum_{\nu=1}^{n} \nu[(1+r)/(1-r)]^2 P_\nu(r,n) \\ &= \frac{k^{(n)}(r)}{n!} = (n+r)/(1-r)^{n+2} . \end{aligned}$$

The general result follows from replacing f by $e^{-i\alpha} f(e^{i\alpha} z)$.

Conversely, the choice $z = 0$ implies $|a_n| \le n$. Therefore the estimates of Theorem B.5 are equivalent to the Bieberbach conjecture.

If the Bieberbach conjecture is true, then the Koebe functions $k(z) = z/(1-\eta z)^2$, $|\eta| = 1$, are extremal functions. They are probably the only ones.

<u>PROBLEM</u>. Assuming that the Bieberbach conjecture is true for a fixed n (or for all n), can one show that the Koebe functions are the only extremal functions?

APPENDIX C. Schiffer's boundary variation and fundamental lemma

A continuum is a closed connected set with more than one point. Let γ be a bounded continuum and D the unbounded component of $\mathbb{C}-\gamma$. Then by the Riemann mapping theorem there exists a conformal mapping φ of D onto $|w|>1$ with $\varphi(\infty)=\infty$ and $\varphi'(\infty)>0$. As $z\to\infty$,

$$\varphi(z) = z/\rho + O(1)$$

for some $\rho>0$. In other words, there exists a conformal mapping ψ of D onto $|w|>\rho$, with

$$\psi(z) = z + O(1)$$

as $z\to\infty$. The number $\rho=\rho(\gamma)$ is the transfinite diameter (exterior mapping radius, capacity) of γ.

LEMMA C.1. Let γ_1 and γ_2 be bounded continua. If $\gamma_1\subset\gamma_2$, then $\rho(\gamma_1)\le\rho(\gamma_2)$.

Proof. $h(w) = 1/[\varphi_1\circ\varphi_2^{-1}(1/w)]$ satisfies the hypotheses of Schwarz's lemma. Therefore

$$1 \ge \lim_{w\to 0}\left|\frac{h(w)}{w}\right| = \lim_{w\to\infty}\left|\frac{w}{\varphi_1\circ\varphi_2^{-1}(w)}\right| = \lim_{z\to\infty}\left|\frac{\varphi_2(z)}{\varphi_1(z)}\right| = \frac{\rho(\gamma_1)}{\rho(\gamma_2)}.$$

LEMMA C.2. If γ is a bounded continuum with diameter d and transfinite diameter ρ, then $2\rho\le d\le 4\rho$ $(\tfrac14 d\le\rho\le\tfrac12 d)$.

Proof. Let $z_1, z_2\in\gamma$ with $|z_1-z_2| = d$. If $\varphi(z) = z/\rho + O(1)$ maps D, the unbounded component of $\mathbb{C}-\gamma$, onto $|w| > 1$, then $\varphi^{-1}(w) = \rho w + O(1)$ maps $|w| > 1$ onto D. Since $h(w) = \rho/[\varphi^{-1}(1/w) - z_2] \in S$ and $h \neq \rho/(z_1 - z_2)$, we have $|\rho/(z_1 - z_2)| = \rho/d \geq \frac{1}{4}$ by Theorem B.3.

On the other hand, $H(w) = [\varphi^{-1}(w) - \varphi^{-1}(-w)]/(2w) = \rho + O(1/w^2)$ is analytic for $|w| > 1$ and finite at ∞. As $|w| \to 1$, the points $\varphi^{-1}(w)$ and $\varphi^{-1}(-w)$ approach γ. Therefore $\limsup_{|w|\to 1} |H(w)| \leq \frac{1}{2}d$. Consequently, $\rho = H(\infty) \leq \frac{1}{2}d$ by the maximum principle.

<u>THEOREM C.3</u> (Schiffer's boundary variation [S2]). Let γ be a continuum and $w_0 \in \gamma$. Then for all sufficiently small $\delta > 0$, there exist $\rho \in \mathbb{R}$ and $A_1 \in \mathbb{C}$ with $\frac{1}{4}\delta \leq \rho \leq \delta$ and $|A_1| \leq 1$, such that functions $F \in H_u(\mathbb{C} - \gamma)$ exist for all $\alpha \in \mathbb{R}$ with

$$F(w) = w + \sum_{n=1}^{\infty} B_n \rho^{n+1} (w - w_0)^{-n} \quad \text{for} \quad |w - w_0| > \delta,$$

where $B_1 = e^{2i\alpha} - A_1$ and $|B_n| \leq 4^{n+1}$ $(n \geq 2)$. Furthermore, if γ is not locally a line segment through w_0, then $|A_1| < 1$.

<u>Proof</u>. Since we have defined continua to be nondegenerate, a $w_1 \in \gamma$ exists with $w_1 \neq w_0$. Now suppose $0 < \delta < |w_1 - w_0|$, and let γ_0 be the component of $\gamma \cap \{|w - w_0| \leq \delta\}$ containing w_0. Necessarily $\gamma_0 \cap \{|w - w_0| = \delta\} \neq \emptyset$ since γ is connected. Let d be the diameter of γ_0 and ρ the transfinite diameter of γ_0. Then $\delta \leq d \leq 2\delta$ by construction, and $\frac{1}{4}\delta \leq \frac{1}{4}d \leq \rho \leq \frac{1}{2}d \leq \delta$ by Lemma C.2.

Let $\varphi(w) = \frac{1}{\rho}w + \beta_0 + \beta_1 w^{-1} + \ldots$ map $\mathbb{C} - \gamma_0$ onto $|\zeta| > 1$. Then $\varphi^{-1}(\zeta) = \rho\zeta - \rho\beta_0 - \beta_1\zeta^{-1} - \ldots$ maps $|\zeta| > 1$ onto $\mathbb{C} - \gamma_0$. Set $\beta_1 = -\rho A_1$. Then $\frac{1}{\rho}\varphi^{-1} \in \Sigma$, and $|A_1| \leq 1$ by Theorem B.1.

Equality occurs iff $\varphi^{-1}(\zeta) = \rho\zeta - \rho\beta_0 - \beta_1\zeta^{-1}$, in which case γ_o is a line segment.

Since the function $\rho(\zeta + e^{2i\alpha}/\zeta)$ maps $|\zeta| > 1$ onto the complement of a segment of length 4ρ and inclination α (cf. Lemma 9.7), the function

$$F = \rho(\varphi + e^{2i\alpha}/\varphi) - \rho\beta_0$$

maps the complement of γ_o onto the complement of a similar segment. For $|w - w_o| > \delta$, the function F has the expansion

$$F(w) = w + \sum_{n=1}^{\infty} B_n \rho^{n+1}(w - w_o)^{-n}$$

where $B_1\rho^2 = \rho\beta_1 + \rho^2 e^{2i\alpha} = \rho^2(e^{2i\alpha} - A_1)$. Finally, since

$$G(\omega) = \frac{1}{\delta} F(w_o + \delta\omega) = \omega + (w_o/\delta) + \sum_{n=1}^{\infty} B_n(\rho/\delta)^{n+1}\omega^{-n}$$

belongs to the class Σ, its coefficients satisfy $|B_n(\rho/\delta)^{n+1}| \le 1/\sqrt{n} \le 1$ by Theorem B.1. Therefore $|B_n| \le (\delta/\rho)^{n+1} \le 4^{n+1}$.

<u>THEOREM C.4</u> (Schiffer's fundamental lemma [S2]). Let $s(w)$ be analytic and nonzero on a bounded continuum γ. Suppose that

$$\Re\{c\, s(w_o)\} \le 0$$

whenever $w_o \in \gamma$ and c is a complex number such that there exists a sequence of functions

$$F_\nu(w) = w + \sum_{n=1}^{\infty} B_{n\nu}\rho_\nu^{n+1}(w - w_o)^{-n}$$

in $H_u(C - \gamma)$ with $\lim_{\nu\to\infty} \rho_\nu = 0$ and $\lim_{\nu\to\infty} B_{1\nu} = c$. Then γ is an analytic arc with some parametric representation $w = w(t)$ that satisfies

$$s(w)\left(\frac{dw}{dt}\right)^2 = 1 .$$

<u>REMARK</u>. We often de-emphasize the particular parametrization of γ chosen above and simply write the differential equation as

$$s(w)(dw)^2 > 0 .$$

Then we say that γ lies on a <u>trajectory</u> of the <u>quadratic differential</u> $s(w)(dw)^2$.

The following is essentially M. Schiffer's original proof as it appears in a set of lecture notes by J. A. Hummel [H9].

<u>Proof</u>. As in the definition on p. 88, we call α a <u>limiting direction</u> of the continuum γ at w_o if points $w_o + \rho_\nu e^{i\alpha_\nu} \in \gamma$ exist with $\lim_{\nu\to\infty} \rho_\nu = 0$ and $\lim_{\nu\to\infty} \alpha_\nu = \alpha$. The continuum γ must have at least one limiting direction at each point, and may have many more.

For simplicity, suppose that $w_o = 0$. Let α be a limiting direction of γ at 0, and $w_\nu = \rho_\nu e^{i\alpha_\nu} \in \gamma$ with $\rho_\nu \to 0$ and $\alpha_\nu \to \alpha$ as $\nu \to \infty$. The functions

$$F_\nu(w) = \frac{-w_\nu}{\log(1 - w_\nu/w)} + \tfrac{1}{2}w_\nu = w - \frac{\rho_\nu^2 e^{2i\alpha_\nu}}{12w} + \ldots$$

have an analytic and univalent branch in $\mathbb{C} - \gamma$, so that

$$\operatorname{Re}\left\{\frac{e^{2i\alpha}}{12}\, s(0)\right\} \geq 0$$

by the hypotheses. Set $s(0) = |s(0)|e^{-2i\tau}$. Then $\operatorname{Re} e^{2i(\alpha-\tau)} \geq 0$

since $s(0) \neq 0$. Hence, every limiting direction α of γ at 0 satisfies either $|\alpha - \tau| \leq \frac{1}{4}\pi$ or $|\alpha - (\tau + \pi)| \leq \frac{1}{4}\pi$ (modulo 2π). It follows that γ must be asymptotically in (at least) one of the two sectors

$$T_1 = \{re^{i\theta}: |\theta - \tau| \leq \tfrac{1}{4}\pi, \ r \geq 0\}$$

$$T_2 = \{re^{i\theta}: |\theta - (\tau + \pi)| \leq \tfrac{1}{4}\pi, \ r \geq 0\}.$$

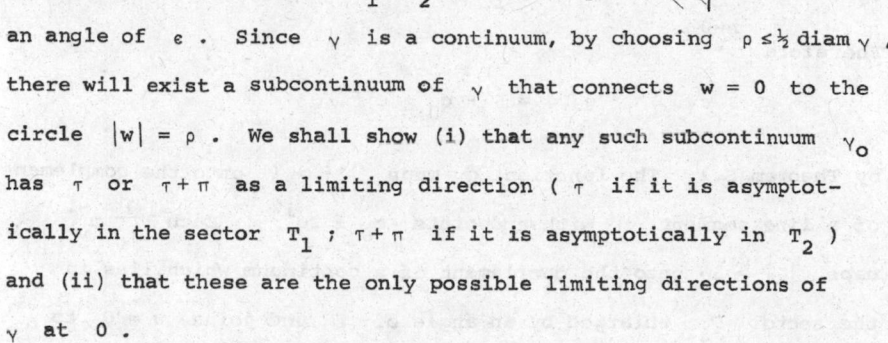

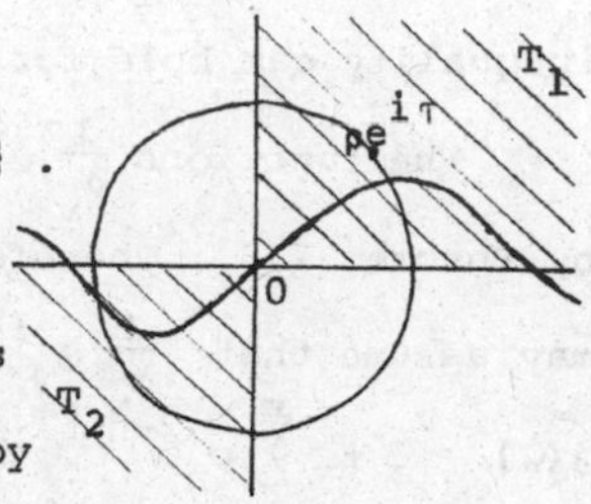

That is, given $\epsilon > 0$, there exists a $\rho > 0$ such that every point of γ in $|w| < \rho$ lies in one of the two sectors T_1, T_2 enlarged by an angle of ϵ. Since γ is a continuum, by choosing $\rho \leq \frac{1}{2}\operatorname{diam}\gamma$, there will exist a subcontinuum of γ that connects $w = 0$ to the circle $|w| = \rho$. We shall show (i) that any such subcontinuum γ_0 has τ or $\tau + \pi$ as a limiting direction (τ if it is asymptotically in the sector T_1 ; $\tau + \pi$ if it is asymptotically in T_2) and (ii) that these are the only possible limiting directions of γ at 0.

(i) Suppose γ_0 is a subcontinuum of γ, that connects $w = 0$ to $|w| = \rho$ in the sector T_1 enlarged by an angle of ϵ. Let $\delta_\nu \to 0$ as $\nu \to \infty$, $0 < \delta_\nu < \rho$, and let γ_ν be a subcontinuum of γ_0 that is contained in $|w| \leq \delta_\nu$ and connects $w = 0$ to $|w| = \delta_\nu$, say at $w_\nu = \delta_\nu e^{i\alpha_\nu}$. From Theorem C.3, there must exist for each ν, a ρ_ν with $\frac{1}{4}\delta_\nu \leq \rho_\nu \leq \delta_\nu$, a conformal mapping

$$\varphi_\nu^{-1}(\zeta) = \rho_\nu(\zeta - \beta_{0\nu} + A_{1\nu}\zeta^{-1} + \dots)$$

of $|\zeta| > 1$ onto $\mathbb{C} - \gamma_\nu$, and for each $\alpha \in \mathbb{R}$, a function

$$F_\nu(w) = w + (e^{2i\alpha} - A_{1\nu})\rho_\nu^2 w^{-1} + \dots$$

in $H_u(C-\gamma_\nu)$. Now $|A_{1\nu}| \le 1$, and by passing to a subsequence, we may assume that $\lim_{\nu\to\infty} A_{1\nu} = A$, $|A| \le 1$. Using the hypothesis with $c = e^{2i\alpha} - A$, we have $\Re\{(e^{2i\alpha}-A)s(0)\} \le 0$, and since $s(0) = |s(0)|e^{-2i\tau}$, we have $\Re\{(e^{2i\alpha}-A)e^{-2i\tau}\} \le 0$. The latter inequality can hold for all α only if $A = e^{2i\tau}$.

The functions $\frac{1}{\rho_\nu}\varphi^{-1}$ belong to the class Σ^0 , which is compact by Theorem 7.5. Therefore, by passing to a further subsequence, we may assume that $\frac{1}{\rho_\nu}\varphi_\nu^{-1} \to g \in \Sigma^0$ as $\nu \to \infty$. If $g(\zeta) = \zeta + \sum_{n=0}^{\infty} c_n\zeta^{-n}$, then $c_1 = A = e^{2i\tau}$, which has modulus one. Therefore

$$g(\zeta) = \zeta + c_0 + e^{2i\tau}/\zeta$$

by Theorem B.1. The function g maps $|\zeta| > 1$ onto the complement of a line segment σ with endpoints $c_0 \pm 2e^{i\tau}$. Each $\frac{1}{\rho_\nu}\varphi_\nu^{-1}$ maps $|\zeta| > 1$ onto the complement of a continuum which lies in the sector T_1 enlarged by an angle of ε and joins $w = 0$ to $w = (\delta_\nu/\rho_\nu)e^{i\alpha_\nu}$. Consequently, σ lies in the sector T_1 and contains $w = 0$, necessarily at an endpoint. For σ to be in T_1 , the point $c_0 - 2e^{i\tau}$ must be the endpoint at $w = 0$, i.e., $c_0 = 2e^{i\tau}$. The other endpoint of σ is $c_0 + 2e^{i\tau} = 4e^{i\tau}$. Since $\delta_\nu e^{i\alpha_\nu}$ is a point on γ_ν at a maximum distance from $w = 0$, it follows that $(\delta_\nu/\rho_\nu)e^{i\alpha_\nu} \to 4e^{i\tau}$ as $\nu \to \infty$. Therefore, $e^{i\alpha_\nu} \to e^{i\tau}$, and τ is a limiting direction of γ_0 at $w = 0$.

A similar argument applies if γ_0 is asymptotically in the sector T_2 . Then $c_0 + 2e^{i\tau}$ is the endpoint of σ at $w=0$ and $e^{i\alpha_\nu} \to -e^{i\tau}$, so that $\tau+\pi$ is a limiting direction of γ_0 at $w=0$.

(ii) Suppose that $\delta_\nu e^{i\beta_\nu} \in \gamma$, $0 < \delta_\nu < \rho$, where $\delta_\nu \to 0$

and $\beta_\nu \to \beta$ as $\nu \to \infty$. Then $|\beta - \tau| \le \frac{1}{4}\pi$ or $|\beta - (\tau+\pi)| \le \frac{1}{4}\pi$ since γ is asymptotically in T_1 or T_2. We shall show that $\beta = \tau$ or $\beta = \tau + \pi$.

From part (i) there is a sequence of points $\delta_\nu e^{i\alpha_\nu} \in \gamma$ with $\alpha_\nu \to \tau$ if $|\beta - \tau| \le \frac{1}{4}\pi$ or $\alpha_\nu \to \tau + \pi$ if $|\beta - (\tau+\pi)| \le \frac{1}{4}\pi$. The functions

$$F_\nu(w) = \frac{\delta_\nu (e^{i\beta_\nu} - e^{i\alpha_\nu})}{\log\left(\dfrac{w - \delta_\nu e^{i\alpha_\nu}}{w - \delta_\nu e^{i\beta_\nu}}\right)} + \tfrac{1}{2}\delta_\nu (e^{i\alpha_\nu} + e^{i\beta_\nu}) = w - \frac{(e^{i\alpha_\nu} - e^{i\beta_\nu})^2}{12w} + \ldots$$

belong to $H_u(C - \gamma)$. Therefore, using $c = -\frac{1}{12}(\lim_{\nu\to\infty} e^{i\alpha_\nu} - e^{i\beta})^2$ in the hypothesis and $s(0) = |s(0)|e^{-2i\tau}$, we have $\Re\{(\lim_{\nu\to\infty} e^{i(\alpha_\nu - \tau)} - e^{i(\beta-\tau)})^2\} \ge 0$. If $|\beta - \tau| \le \frac{1}{4}\pi$, then $0 \ge [\cos(\beta - \tau)][\cos(\beta - \tau) - 1] = \Re(1 - e^{i(\beta-\tau)})^2 \ge 0$; hence $\beta = \tau$. If $|\beta - (\tau+\pi)| \le \frac{1}{4}\pi$, then $0 \ge [\cos(\beta - (\tau+\pi))][\cos(\beta - (\tau+\pi)) - 1] = \Re(-1 - e^{i(\beta-\tau)})^2 \ge 0$; hence $\beta = \tau + \pi$. Consequently, τ and $\tau + \pi$ are the only possible limiting directions of γ at $w = 0$.

The same reasoning holds at any point $w \in \gamma$. That is, if $s(w) = |s(w)|e^{-2i\tau(w)}$, then the only possible limiting directions of γ at w are $\tau(w)$ and $\tau(w) + \pi$.

Let Δ be a neighborhood of $w_0 \in \gamma$, small enough that $W(w) = \int_{w_0}^{w} [s(\omega)]^{\frac{1}{2}} d\omega$ is univalent in Δ. Let $\Gamma = W(\gamma \cap \Delta)$. At any point $w \in \gamma \cap \Delta$,

$$\frac{dW}{dw} = |s(w)|^{\frac{1}{2}} e^{-\frac{1}{2}i\tau(w)},$$

so that the only possible limiting directions of Γ at any point are 0 and π. By a result of U. S. Haslam-Jones [H1]

(Theorem C.5 below), it follows that Γ is a horizontal line segment. Since $W(w_o) = 0$, we may parametrize Γ by $W = t$, $a \le t \le b$. Then γ is given in the neighborhood Δ by $w = W^{-1}(t)$. Hence, γ is an analytic arc satisfying $\int_{w_o}^{w} [s(\omega)]^{\frac{1}{2}} d\omega = t$, at least in Δ. Thus, γ is an integral curve of the differential equation

$$s(w)\left(\frac{dw}{dt}\right)^2 = 1 .$$

Since w_o is arbitrary, this completes the proof.

Observe that if $s(w) = 0$ at some point, then γ may branch or become nonanalytic at that point.

<u>EXERCISE</u>. If $s(w)$ has a zero of order n on γ, show that γ will have locally at most $n+2$ analytic arcs meeting at that point.

In the proof of Theorem C.4, we used a result of U. S. Haslam-Jones [H1]. He actually proved a more general result; however, we shall include a proof for this special case. It is given in some detail since the theorem seems to invite incomplete proofs. For the following proof we are indebted to J. A. Hummel.

<u>THEOREM C.5</u> (Haslam-Jones [H1]). Let E be a bounded continuum that has only horizontal limiting directions at every point. Then E is a horizontal line segment.

<u>Proof</u>. In the proof we denote the real and imaginary parts of a complex number z by x and y, e.g., $z_o = x_o + iy_o$. We also denote the projection of a set A onto the y-axis by A' and

(one-dimensional) Lebesgue measure on the y-axis by m .

Since E is a continuum, its projection E' is a closed line segment. Therefore it is sufficient to prove that $m(E') = 0$. We proceed by contradiction, assuming $m(E') > 0$.

For any $z_o \in E$, $\epsilon > 0$, and integer $n > 0$, define

$$S_n^\epsilon(z_o) = \{z\colon 0 < |z - z_o| < 1/n \ , \ |y - y_o| > \epsilon |x - x_o|\} .$$

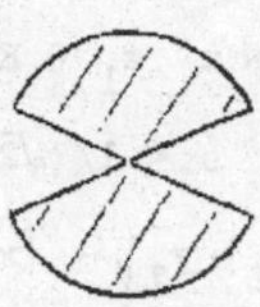

Since E has only horizontal limiting directions, there exists for each ϵ and z , an n such that $S_n^\epsilon(z) \cap E = \emptyset$. In particular, if

$$B_n = \{z \in E\colon S_n^1(z) \cap E = \emptyset\} ,$$

then $\bigcup_{n=1}^{\infty} B_n = E$.

Suppose $\{z_k\} \subset B_n$ and $z_k \to z_o$ as $k \to \infty$, but $z_o \notin B_n$. Then there exists a point $\zeta \in S_n^1(z_o) \cap E$. Since $S_n^1(z_o)$ is open, there also exists a $\rho > 0$ such that the disk $\{z\colon |z - \zeta| < \rho\} \subset S_n^1(z_o)$. Thus, if $|z_k - z_o| < \rho$, then $\zeta \in S_n^1(z_k)$. This contradicts the assumption that $z_k \in B_n$. Consequently, each B_n is a closed subset of E , hence compact.

Since $m(\bigcup_{n=1}^{\infty} B_n') = m(E') > 0$, there exists an n_o such that $m(B_{n_o}') > 0$. For this n_o and any $z_o \in B_{n_o}$, let

$$Q(z_o) = \{z\colon |x - x_o| < \rho_o \ , \ |y - y_o| < \rho_o\} \quad \text{where} \quad \rho_o = 1/(4n_o) .$$

Since B_{n_o} is compact, finitely many of these squares cover B_{n_o} , say $B_{n_o} \subset \bigcup_{k=1}^{N} Q(z_k)$. Since $0 < m(B_{n_o}') \le m(\bigcup_{k=1}^{N} Q(z_k)')$, there exists a square $Q_o = Q(z_k)$ such that if we set $E_o = B_{n_o} \cap \overline{Q}_o$, then $m(E_o') > 0$.

Suppose $z_1, z_2 \in E_o$ and $z_1 \neq z_2$. Then $|z_1 - z_2| \le 2\sqrt{2}\,\rho_o$. Since $z_1 \in B_{n_o}$ and $z_2 \in E$, we have $z_2 \notin S^1_{n_o}(z_1)$; in particular, $|y_2 - y_1| \le |x_2 - x_1|$. It follows that E_o is a graph over its projection on the x-axis.

Fix $\varepsilon = \tfrac{1}{2} n_o m(E_o') > 0$. For any $\rho > 0$ and $z_1 \in E_o$, define

$$R^{\varepsilon}_{\rho}(z_1) = \{z\colon\ |x - x_1| \le \rho\ ,\ |y - y_1| \le \varepsilon\rho\} .$$

Since $z_1 \in E_o$, there exists an n_1 such that $S^{\varepsilon}_{n_1}(z_1) \cap E_o = \phi$. If $\rho < 1/(\sqrt{2}\, n_1)$ and $z_2 \in E_o$ is such that $|x_2 - x_1| < \rho$, then $|y_2 - y_1| \le |x_2 - x_1| < \rho$ and $|z_1 - z_2| < \sqrt{2}\,\rho < 1/n_1$. Since $z_2 \notin S^{\varepsilon}_{n_1}(z_1)$, it follows that $z_2 \in R^{\varepsilon}_{\rho}(z_1)$. If $z_1 \in E_o$, we have shown that there exists a ρ_1 $(= 1/(\sqrt{2}\, n_1))$ such that $\{z \in E_o\colon |x - x_1| < \rho\} \subset R^{\varepsilon}_{\rho}(z_1)$ for all ρ, $0 < \rho < \rho_1$.

The set of all such $R^{\varepsilon}_{\rho}(z_1)'$ forms a Vitali covering of E_o'. Hence, by the Vitali covering theorem, there exist $R_j = R^{\varepsilon}_{\rho_j}(z_j)$, $j = 1,\ldots,N$, with each $z_j \in E_o$, such that the R_j' are disjoint closed intervals and $\sum_{j=1}^{N} m(R_j') > \tfrac{1}{2} m(E_o')$.

Since E_o is a graph over its projection on the x-axis and since the R_j' are disjoint, we may index the z_j so that $x_1 < x_2 < \ldots < x_N$ where $x_{j+1} - x_j > \rho_{j+1} + \rho_j$ $(j = 1,\ldots,N-1)$. Since $E_o \subset \overline{Q}_o$, we have

$$1/(2n_o) = 2\rho_o \ge x_N - x_1 = \sum_{j=1}^{N-1} (x_{j+1} - x_j) > \sum_{j=1}^{N} \rho_j$$

$$= \frac{1}{2\varepsilon} \sum_{j=1}^{N} 2\varepsilon\rho_j = \frac{1}{n_o m(E_o')} \sum_{j=1}^{N} m(R_j') > 1/(2n_o) .$$

From this contradiction the theorem follows.

REFERENCES

A 1. S. Agmon, A. Douglis, and L. Nirenberg: Estimates near the boundary for solutions of elliptic partial differential equations satisfying general boundary conditions I, Comm. Pure Appl. Math. 12 (1959), 623 - 727.

2. D. Aharonov: On the Bieberbach conjecture for functions with a small second coefficient, Israel J. Math. 15 (1973), 137 - 139.

3. D. Aharonov and S. Friedland: On an inequality connected with the coefficient conjecture for functions of bounded boundary rotation, Ann. Acad. Sci. Fenn. AI 524 (1972).

4. L. V. Ahlfors: Quasiconformal reflections, Acta Math. 109 (1963), 291 - 301.

5. L. V. Ahlfors: Lectures on Quasiconformal Mappings, Van Nostrand Math. Studies 10, New York, 1966.

6. L. V. Ahlfors: Conformal Invariants, McGraw - Hill, New York, 1973.

7. L. V. Ahlfors: Sufficient conditions for quasiconformal extension, Princeton Annals of Math. Studies 79 (1974), 23 - 29.

8. L. V. Ahlfors and H. Grunsky: Über die Blochsche Konstante, Math. Z. 42 (1937), 671 - 673.

9. E. M. Alfsen: Compact Convex Sets and Boundary Integrals, Ergebnisse der Math. 57, Springer-Verlag, New York-Heidelberg-Berlin, 1971.

B 1. P. P. Belinskiĭ: Solution of extremum problems in the theory of quasiconformal mappings by variational methods (Russ.), Sibirsk. Mat. Ž. 1 (1960), 303 - 330.

2. L. Bieberbach: Über die Koeffizienten derjenigen Potenzreihen, welche eine schlichte Abbildung des Einheitskreises vermitteln, S.-B. Preuss. Akad. Wiss. 1916, 940 - 955.

3. P. A. Biluta: On an extremal problem for quasiconformal mappings of finitely connected domains, Siberian Math. J. 13 (1972), 16 - 22.

4. P. A. Biluta and S. L. Kruškal': On extremal quasiconformal mappings, Soviet Math. Dokl. 12 (1971), 76 - 79.

B 5. A. Bloch: Les théorèmes de M. Valiron sur les fonctions entières et la théorie de l'uniformisation, Ann. Fac. Sci. Univ. Toulouse 17 (1925), 1 - 22.

6. N. Bourbaki: Éléments de Mathématique, Livre V, Espaces vectoriels topologiques, Hermann et Cie, Act. Sci. et Ind. 1189, 1129, Paris, 1953, 1955.

7. D. A. Brannan: On coefficient problems for certain power series, London Math. Soc. Lecture Note Series 12 (1974), 17 - 27.

8. D. A. Brannan, J. G. Clunie, and W. E. Kirwan: On the coefficient problem for functions of bounded boundary rotation, Ann. Acad. Sci. Fenn. AI 523 (1973).

9. L. Brickman: Extreme points of the set of univalent functions, Bull. Amer. Math. Soc. 76 (1970), 372 - 374.

10. L. Brickman, T. H. MacGregor, and D. R. Wilken: Convex hulls of some classical families of univalent functions, Trans. Amer. Math. Soc. 156 (1971), 91 - 107.

11. L. Brickman and D. R. Wilken: Support points of the set of univalent functions, Proc. Amer. Math. Soc. 42 (1974), 523 - 528.

C 1. R. Caccioppoli: Sui funzionali lineari nel campo delle funzioni analitiche, Atti Accad. Naz. Lincei Rend. Cl. Sci. Fis. Mat. Natur. 13 (1931), 263 - 266.

2. Z. Charzynski and M. Schiffer: A geometric proof of the Bieberbach conjecture for the fourth coefficient, Scripta Math. 25 (1960), 173 - 181.

3. Z. Charzynski and M. Schiffer: A new proof of the Bieberbach conjecture for the fourth coefficient, Arch. Rational Mech. Anal. 5 (1960), 187 - 193.

4. J. Curtiss: Faber polynomials and the Faber series, Amer. Math. Monthly 78 (1971), 577 - 596.

D 1. N. Dunford and J. T. Schwartz: Linear Operators Part I: General Theory, Interscience, New York, 1958.

F 1. C. FitzGerald: Quadratic inequalities and coefficient estimates for schlicht functions, Arch. Rational Mech. Anal. 46 (1972), 356 - 368.

2. S. Friedland: Generalized Hadamard inequality and its applications, Linear and Multilinear Algebra, to appear.

G 1. P. R. Garabedian, G. Ross, and M. Schiffer: On the Bieberbach conjecture for even n, J. Math. Mech. 14 (1965), 975 - 989.

2. P. R. Garabedian and M. Schiffer: A coefficient inequality for schlicht functions, Ann. of Math. 61 (1955), 116 - 136.

3. P. R. Garabedian and M. Schiffer: A proof of the Bieberbach conjecture for the fourth coefficient, J. Rational Mech. Anal. 4 (1955), 427 - 465.

4. P. R. Garabedian and M. Schiffer: The local maximum theorem for the coefficients of univalent functions, Arch. Rational Mech. Anal. 26 (1967), 1 - 32.

5. F. W. Gehring: Definitions for a class of plane quasiconformal mappings, Nagoya Math. J. 29 (1967), 175 - 184.

6. F. W. Gehring and O. Lehto: On the total differentiability of functions of a complex variable, Ann. Acad. Sci. Fenn. AI 272 (1959).

7. G. M. Goluzin: Method of variations in the theory of conformal representation, Mat. Sbornik 19 (1946), 203 - 236, and 21 (1947), 83 - 117.

8. G. M. Goluzin: Geometric Theory of Functions of a Complex Variable, Amer. Math. Soc. Transl. of Math. Monographs 29, Providence, RI, 1969.

9. T. H. Gronwall: Some remarks on conformal representation, Ann. of Math. 16 (1914/15), 72 - 76.

10. H. Grunsky: Neue Abschätzungen zur konformen Abbildung ein- und mehrfach zusammenhängender Bereiche, Schr. Math. Inst. u. Inst. Angew. Math. Univ. Berlin 1 (1932), 95 - 140.

11. H. Grunsky: Koeffizientenbedingungen für schlicht abbildende meromorphe Funktionen, Math. Z. 45 (1939), 29 - 61.

H 1. U. S. Haslam - Jones: Tangential properties of a plane set of points, Quart. J. Math. 7 (1936), 116 - 123.

2. W. Hayman: The asymptotic behaviour of p-valent functions, Proc. London Math. Soc. 5 (1955), 257 - 284.

3. W. Hengartner and G. Schober: A remark on level curves for domains convex in one direction, Applicable Anal. 3 (1973), 101 - 106.

H 4. W. Hengartner and G. Schober: Extreme points for some classes of univalent functions, Trans. Amer. Math. Soc. 185 (1973), 265 - 270.

5. W. Hengartner and G. Schober: Compact families of univalent functions and their support points, Mich. Math. J. 21(1974), 205-217.

6. W. Hengartner and G. Schober: Propriétés des points d'appui des familles compactes de fonctions univalentes, C. R. Acad. Sci. Paris Sér. A 279 (1974), 551 - 553.

7. J. Hocking and G. Young: Topology, Addison - Wesley, Reading, Mass., 1961.

8. F. Holland: The extreme points of a class of functions with positive real part, Math. Ann. 202 (1973), 85 - 87.

9. J. A. Hummel: Lectures on Variational Methods in the Theory of Univalent Functions, Univ. of Maryland Lecture Notes, 1970.

I 1. L. P. Il'ina: Estimates for the coefficients of univalent functions in terms of the second coefficient, Math. Notes 13 (1973), 215 - 218.

J 1. J. A. Jenkins: Univalent Functions and Conformal Mapping, Ergebnisse der Math. 18, Springer-Verlag, Berlin-Heidelberg-New York, 1965.

2. J. A. Jenkins: On an inequality considered by Robertson, Proc. Amer. Math. Soc. 19 (1968), 549 - 550.

3. G. Julia: Sur une équation aux dérivées fonctionelles liée à la représentation conforme, Ann. École Norm. Sup. 39 (1922), 1 - 28.

K 1. W. E. Kirwan and G. Schober: On extreme points and support points for some families of univalent functions, Duke Math. J., to appear.

2. P. Koebe: Über die Uniformisierung beliebiger analytischen Kurven, Nachr. Ges. Wiss. Göttingen 2 (1907), 191 - 210.

3. P. Koebe: Abhandlungen zur Theorie der konforme Abbildung VI, Math. Z. 7 (1920), 235 - 301.

4. P. Koebe: Über die konforme Abbildung endlich- und unendlich-vielfach zusammenhängender symmetrischer Bereiche, Acta Math. 43 (1922), 263 - 287.

K 5. G. Köthe: Topological Vector Spaces I, Grundlehren math. Wiss. 159, Springer-Verlag, New York, 1969.

6. W. Kraus: Über den Zusammenhang einiger Charakteristiken eines einfach zusammenhängenden Bereiches mit der Kreisabbildung, Mitt. Math. Sem. Giessen 21 (1932), 1 - 28.

7. S. L. Kruškal': Some extremal problems for conformal and quasiconformal mappings, and On the connection between variational problems for conformal and quasiconformal mappings, Siberian Math. J. 12 (1971), 541 - 559 and 769 - 776.

8. R. Kühnau: Wertannahmeprobleme bei quasikonformen Abbildungen mit ortsabhängiger Dilatationsbeschränkung, Math. Nachr. 40 (1969), 1 - 11.

9. R. Kühnau: Verzerrungssätze und Koeffizientenbedingungen vom Grunskyschen Typ für quasikonforme Abbildungen, Math. Nach. 48 (1971), 77 - 105.

10. R. Kühnau: Eine funktionentheoretische Randwertaufgabe in der Theorie der quasikonformen Abbildungen, Indiana Univ. Math. J. 21 (1971/72), 1 - 10.

11. R. Kühnau: Zum Koeffizientenproblem bei den quasikonform fortsetzbaren schlichten konformen Abbildungen, Math. Nachr. 55 (1973), 225 - 231.

12. R. Kühnau: Zur analytischen Darstellung gewisser Extremalfunktionen der quasikonformen Abbildung, Math. Nachr. 60 (1974), 53 - 62.

L 1. E. Landau: Einige Bemerkungen über schlichte Abbildung, Jber. Deutsch. Math. - Verein 34 (1925/26), 239 - 243.

2. E. Landau: Der Picard - Schottkysche Satz und die Blochsche Konstante, S. - B. Preuss. Akad. Wiss. 1926, 467 - 474.

3. O. Lehto: Schlicht functions with a quasiconformal extension, Ann. Acad. Sci. Fenn. AI 500 (1971).

4. O. Lehto: Conformal Mappings and Teichmüller Spaces, Technion Lecture Notes, Israel Inst. Tech., Haifa, 1973.

5. O. Lehto and K. I. Virtanen: Quasiconformal Mappings in the Plane, Grundlehren math. Wiss. 126, Springer-Verlag, New York-Heidelberg-Berlin, 1973.

6. K. Löwner: Untersuchungen über schlichte konforme Abbildungen des Einheitskreises I, Math. Ann. 89 (1923), 103 - 121.

M 1. A. Marx: Untersuchungen über schlichte Abbildungen, Math. Ann. 107 (1932/33), 40 - 67.

2. J. O. McLeavey: Extremal problems in classes of analytic univalent functions with quasiconformal extensions, Trans. Amer. Math. Soc. 195 (1974), 327 - 343.

O 1. M. Ozawa: On the Bieberbach conjecture for the sixth coefficient, Kodai Math. Sem. Rep. 21 (1969), 97 - 128.

P 1. V. Paatero: Über die konforme Abbildung von Gebieten deren Ränder von beschränkter Drehung sind, and Über Gebiete von beschränkter Randdrehung, Ann. Acad. Sci. Fenn. A 33 (1931) and 37 (1933).

2. R. Pederson: A proof of the Bieberbach conjecture for the sixth coefficient, Arch. Rational Mech. Anal. 31 (1968), 331 - 351.

3. R. Pederson and M. Schiffer: A proof of the Bieberbach conjecture for the fifth coefficient, Arch. Rational Mech. Anal. 45 (1972), 161 - 193.

4. A. Pfluger: Lineare Extremalprobleme bei schlichten Funktionen, Ann. Acad. Sci. Fenn. AI 489 (1971).

5. R. R. Phelps: Lectures on Choquet's Theorem, Van Nostrand Math. Studies 7, New York, 1966.

6. G. Pólya and I. J. Schoenberg: Remarks on de la Vallée Poussin means and convex conformal maps of the circle, Pacific J. Math. 8 (1958), 295 - 334.

7. Chr. Pommerenke: On a variational method for univalent functions, Mich. Math. J. 17 (1970), 1 - 3.

8. Chr. Pommerenke: Univalent Functions, Vandenhoeck & Ruprecht, Göttingen, 1975.

R 1. H. Renelt: Modifizierung und Erweiterung einer Schifferschen Variationsmethode für quasikonforme Abbildungen, Math. Nachr. 55 (1973), 353 - 379.

2. M. S. Robertson: A generalization of the Bieberbach coefficient problem for univalent functions, Mich. Math. J. 13 (1966), 185 - 192.

R 3. M. S. Robertson: Quasi-subordination and coefficient conjectures, Bull. Amer. Math. Soc. 76 (1970), 1 - 9.

4. St. Ruscheweyh and T. Sheil - Small: Hadamard products of schlicht functions and the Pólya - Schoenberg conjecture, Comment. Math. Helv. 48 (1973), 119 - 135.

S 1. A. C. Schaeffer and D. C. Spencer: Coefficient Regions for Schlicht Functions, Amer. Math. Soc. Colloq. Publ. 35, Providence, RI, 1950.

2. M. Schiffer: A method of variation within the family of simple functions, Proc. London Math. Soc. 44 (1938), 432 - 449.

3. M. Schiffer: On the coefficients of simple functions, Proc. London Math. Soc. 44 (1938), 450 - 452.

4. M. Schiffer: Sur un problème d'extrémum de la représentation conforme, Bull. Soc. Math. France 66 (1938), 48 - 55.

5. M. Schiffer: Variation of the Green function and theory of the p - valued functions, Amer. J. Math. 65 (1943), 341 - 360.

6. M. Schiffer: Faber polynomials in the theory of univalent functions, Bull. Amer. Math. Soc. 54 (1948), 503 - 517.

7. M. Schiffer: Fredholm eigenvalues of multiply connected domains, Pacific J. Math. 9 (1959), 211 - 269.

8. M. Schiffer: Extremum problems and variational methods in conformal mapping, Proc. International Congress Mathematicians 1958, Cambridge Univ. Press (1960), 211 - 231.

9. M. Schiffer: Fredholm eigenvalues and conformal mapping, Rend. Mat. 22 (1963), 447 - 468.

10. M. Schiffer: A variational method for univalent quasiconformal mappings, Duke Math. J. 33 (1966), 395 - 412.

11. M. Schiffer and G. Schober: An extremal problem for the Fredholm eigenvalues, Arch. Rational Mech. Anal. 44 (1971), 83 - 92 and 46 (1972), 394.

12. M. Schiffer and G. Schober: Coefficient problems and generalized Grunsky inequalities for schlicht functions with quasiconformal extensions, Arch. Rational Mech. Anal., to appear.

S 13. I. Schur: Bemerkungen zur Theorie der beschränkten Bilinearformen mit unendlich vielen Veränderlichen, J. Reine Angew. Math. 140 (1911), 1 - 28.

14. I. Schur: Über Potenzreihen, die im Innern des Einheitskreises beschränkt sind, J. Reine Angew. Math. 147 (1917), 205 - 232, and 148 (1918), 122 - 145.

15. I. Schur: Ein Satz über quadratische Formen mit komplexen Koeffizienten, Amer. J. Math. 67 (1945), 472 - 480.

16. R. J. Sibner: Remarks on the Koebe Kreisnormierungsproblem, Comment. Math. Helv. 43 (1968), 289 - 295.

17. G. Springer: Extreme Punkte der konvexen Hülle schlichter Funktionen, Math. Ann. 129 (1955), 230 - 232.

18. G. Springer: Fredholm eigenvalues and quasiconformal mapping, Acta Math. 111 (1964), 121 - 142.

19. K. Strebel: Über das Kreisnormierungsproblem der konformen Abbildung, Ann. Acad. Sci. Fenn. AI 101 (1951).

20. K. Strebel: Ein Konvergensatz für Folgen quasikonformer Abbildungen, Comment. Math. Helv. 44 (1969), 469 - 475.

21. E. Strohäcker: Beitrage zur Theorie der schlichten Funktionen, Math. Z. 37 (1933), 356 - 380.

22. T. J. Suffridge: Convolutions of convex functions, J. Math. Mech. 15 (1966), 795 - 804.

23. T. J. Suffridge: Some remarks on convex maps of the unit disc, Duke Math. J. 37 (1970), 775 - 777.

T 1. O. Toeplitz: Die linearen vollkommenen Räume der Funktionentheorie, Comment. Math. Helv. 23 (1949), 222 - 242.

V 1. J. Väisälä: Lectures on n - Dimensional Quasiconformal Mappings, Springer-Verlag Lecture Notes in Math. 229, Berlin-Heidelberg-New York, 1971.

W 1. S. E. Warschawski: On Hadamard's variational formula for Green's function, J. Math. Mech. 9 (1960), 497 - 512.

INDEX

Vol. 309: D. H. Sattinger, Topics in Stability and Bifurcation Theory. VI, 190 pages. 1973. DM 20,-

Vol. 310: B. Iversen, Generic Local Structure of the Morphisms in Commutative Algebra. IV, 108 pages. 1973. DM 18,-

Vol. 311: Conference on Commutative Algebra. Edited by J. W. Brewer and E. A. Rutter. VII, 251 pages. 1973. DM 24,-

Vol. 312: Symposium on Ordinary Differential Equations. Edited by W. A. Harris, Jr. and Y. Sibuya. VIII, 204 pages. 1973. DM 22,-

Vol. 313: K. Jörgens and J. Weidmann, Spectral Properties of Hamiltonian Operators. III, 140 pages. 1973. DM 18,-

Vol. 314: M. Deuring, Lectures on the Theory of Algebraic Functions of One Variable. VI, 151 pages. 1973. DM 18,-

Vol. 315: K. Bichteler, Integration Theory (with Special Attention to Vector Measures). VI, 357 pages. 1973. DM 29,-

Vol. 316: Symposium on Non-Well-Posed Problems and Logarithmic Convexity. Edited by R. J. Knops. V, 176 pages. 1973. DM 20,-

Vol. 317: Séminaire Bourbaki – vol. 1971/72. Exposés 400–417. IV, 361 pages. 1973. DM 29,-

Vol. 318: Recent Advances in Topological Dynamics. Edited by A. Beck. VIII, 285 pages. 1973. DM 27,-

Vol. 319: Conference on Group Theory. Edited by R. W. Gatterdam and K. W. Weston. V, 188 pages. 1973. DM 20,-

Vol. 320: Modular Functions of One Variable I. Edited by W. Kuyk. V, 195 pages. 1973. DM 20,-

Vol. 321: Séminaire de Probabilités VII. Edité par P. A. Meyer. VI, 322 pages. 1973. DM 29,-

Vol. 322: Nonlinear Problems in the Physical Sciences and Biology. Edited by I. Stakgold, D. D. Joseph and D. H. Sattinger. VIII, 357 pages. 1973. DM 29,-

Vol. 323: J. L. Lions, Perturbations Singulières dans les Problèmes aux Limites et en Contrôle Optimal. XII, 645 pages. 1973. DM 46,-

Vol. 324: K. Kreith, Oscillation Theory. VI, 109 pages. 1973. DM 18,-

Vol. 325: C.-C. Chou, La Transformation de Fourier Complexe et L'Equation de Convolution. IX, 137 pages. 1973. DM 18,-

Vol. 326: A. Robert, Elliptic Curves. VIII, 264 pages. 1973. DM 24,-

Vol. 327: E. Matlis, One-Dimensional Cohen-Macaulay Rings. XII, 157 pages. 1973. DM 20,-

Vol. 328: J. R. Büchi and D. Siefkes, The Monadic Second Order Theory of All Countable Ordinals. VI, 217 pages. 1973. DM 22,-

Vol. 329: W. Trebels, Multipliers for (C, α)-Bounded Fourier Expansions in Banach Spaces and Approximation Theory. VII, 103 pages. 1973. DM 18,-

Vol. 330: Proceedings of the Second Japan-USSR Symposium on Probability Theory. Edited by G. Maruyama and Yu. V. Prokhorov. VI, 550 pages. 1973. DM 40,-

Vol. 331: Summer School on Topological Vector Spaces. Edited by L. Waelbroeck. VI, 226 pages. 1973. DM 22,-

Vol. 332: Séminaire Pierre Lelong (Analyse) Année 1971-1972. V, 131 pages. 1973. DM 18,-

Vol. 333: Numerische, insbesondere approximationstheoretische Behandlung von Funktionalgleichungen. Herausgegeben von R. Ansorge und W. Törnig. VI, 296 Seiten. 1973. DM 27,-

Vol. 334: F. Schweiger, The Metrical Theory of Jacobi-Perron Algorithm. V, 111 pages. 1973. DM 18,-

Vol. 335: H. Huck, R. Roitzsch, U. Simon, W. Vortisch, R. Walden, B. Wegner und W. Wendland, Beweismethoden der Differentialgeometrie im Großen. IX, 159 Seiten. 1973. DM 20,-

Vol. 336: L'Analyse Harmonique dans le Domaine Complexe. Edité par E. J. Akutowicz. VIII, 169 pages. 1973. DM 20,-

Vol. 337: Cambridge Summer School in Mathematical Logic. Edited by A. R. D. Mathias and H. Rogers. IX, 660 pages. 1973. DM 46,-

Vol. 338: J. Lindenstrauss and L. Tzafriri, Classical Banach Spaces. IX, 243 pages. 1973. DM 24,-

Vol. 339: G. Kempf, F. Knudsen, D. Mumford and B. Saint-Donat, Toroidal Embeddings I. VIII, 209 pages. 1973. DM 22,-

Vol. 340: Groupes de Monodromie en Géométrie Algébrique. (SGA 7 II). Par P. Deligne et N. Katz. X, 438 pages. 1973. DM 44,-

Vol. 341: Algebraic K-Theory I, Higher K-Theories. Edited by H. Bass. XV, 335 pages. 1973. DM 29,-

Vol. 342: Algebraic K-Theory II, "Classical" Algebraic K-Theory, and Connections with Arithmetic. Edited by H. Bass. XV, 527 pages. 1973. DM 40,-

Vol. 343: Algebraic K-Theory III, Hermitian K-Theory and Geometric Applications. Edited by H. Bass. XV, 572 pages. 1973. DM 40,-

Vol. 344: A. S. Troelstra (Editor), Metamathematical Investigation of Intuitionistic Arithmetic and Analysis. XVII, 485 pages. 1973. DM 38,-

Vol. 345: Proceedings of a Conference on Operator Theory. Edited by P. A. Fillmore. VI, 228 pages. 1973. DM 22,-

Vol. 346: Fučík et al., Spectral Analysis of Nonlinear Operators. II, 287 pages. 1973. DM 26,-

Vol. 347: J. M. Boardman and R. M. Vogt, Homotopy Invariant Algebraic Structures on Topological Spaces. X, 257 pages. 1973. DM 24,-

Vol. 348: A. M. Mathai and R. K. Saxena, Generalized Hypergeometric Functions with Applications in Statistics and Physical Sciences. VII, 314 pages. 1973. DM 26,-

Vol. 349: Modular Functions of One Variable II. Edited by W. Kuyk and P. Deligne. V, 598 pages. 1973. DM 38,-

Vol. 350: Modular Functions of One Variable III. Edited by W. Kuyk and J.-P. Serre. V, 350 pages. 1973. DM 26,-

Vol. 351: H. Tachikawa, Quasi-Frobenius Rings and Generalizations. XI, 172 pages. 1973. DM 20,-

Vol. 352: J. D. Fay, Theta Functions on Riemann Surfaces. V, 137 pages. 1973. DM 18,-

Vol. 353: Proceedings of the Conference on Orders, Group Rings and Related Topics. Organized by J. S. Hsia, M. L. Madan and T. G. Ralley. X, 224 pages. 1973. DM 22,-

Vol. 354: K. J. Devlin, Aspects of Constructibility. XII, 240 pages. 1973. DM 24,-

Vol. 355: M. Sion, A Theory of Semigroup Valued Measures. V, 140 pages. 1973. DM 18,-

Vol. 356: W. L. J. van der Kallen, Infinitesimally Central Extensions of Chevalley Groups. VII, 147 pages. 1973. DM 18,-

Vol. 357: W. Borho, P. Gabriel und R. Rentschler, Primideale in Einhüllenden auflösbarer Lie-Algebren. V, 182 Seiten. 1973. DM 20,-

Vol. 358: F. L. Williams, Tensor Products of Principal Series Representations. VI, 132 pages. 1973. DM 18,-

Vol. 359: U. Stammbach, Homology in Group Theory. VIII, 183 pages. 1973. DM 20,-

Vol. 360: W. J. Padgett and R. L. Taylor, Laws of Large Numbers for Normed Linear Spaces and Certain Fréchet Spaces. VI, 111 pages. 1973. DM 18,-

Vol. 361: J. W. Schutz, Foundations of Special Relativity: Kinematic Axioms for Minkowski Space Time. XX, 314 pages. 1973. DM 26,-

Vol. 362: Proceedings of the Conference on Numerical Solution of Ordinary Differential Equations. Edited by D. Bettis. VIII, 490 pages. 1974. DM 34,-

Vol. 363: Conference on the Numerical Solution of Differential Equations. Edited by G. A. Watson. IX, 221 pages. 1974. DM 20,-

Vol. 364: Proceedings on Infinite Dimensional Holomorphy. Edited by T. L. Hayden and T. J. Suffridge. VII, 212 pages. 1974. DM 20,-

Vol. 365: R. P. Gilbert, Constructive Methods for Elliptic Equations. VII, 397 pages. 1974. DM 26,-

Vol. 366: R. Steinberg, Conjugacy Classes in Algebraic Groups (Notes by V. V. Deodhar). VI, 159 pages. 1974. DM 18,-

Vol. 367: K. Langmann und W. Lütkebohmert, Cousinverteilungen und Fortsetzungssätze. VI, 151 Seiten. 1974. DM 16,-

Vol. 368: R. J. Milgram, Unstable Homotopy from the Stable Point of View. V, 109 pages. 1974. DM 16,-

Vol. 369: Victoria Symposium on Nonstandard Analysis. Edited by A. Hurd and P. Loeb. XVIII, 339 pages. 1974. DM 26,-

Vol. 370: B. Mazur and W. Messing, Universal Extensions and One Dimensional Crystalline Cohomology. VII, 134 pages. 1974. DM 16,–

Vol 371: V. Poenaru, Analyse Différentielle. V, 228 pages 1974 DM 20,–

Vol. 372: Proceedings of the Second International Conference on the Theory of Groups 1973. Edited by M. F. Newman. VII, 740 pages. 1974. DM 48,–

Vol. 373: A. E. R. Woodcock and T. Poston, A Geometrical Study of the Elementary Catastrophes. V, 257 pages. 1974. DM 22,–

Vol. 374: S. Yamamuro, Differential Calculus in Topological Linear Spaces. IV, 179 pages. 1974. DM 18,–

Vol. 375: Topology Conference 1973. Edited by R. F. Dickman Jr. and P. Fletcher. X, 283 pages. 1974. DM 24,–

Vol. 376: D. B. Osteyee and I. J. Good, Information, Weight of Evidence, the Singularity between Probability Measures and Signal Detection. XI, 156 pages. 1974. DM 16.–

Vol. 377: A. M. Fink, Almost Periodic Differential Equations VIII, 336 pages. 1974. DM 26,–

Vol. 378: TOPO 72 – General Topology and its Applications. Proceedings 1972. Edited by R. Alò, R. W. Heath and J. Nagata. XIV, 651 pages. 1974. DM 50,–

Vol. 379: A. Badrikian et S. Chevet, Mesures Cylindriques, Espaces de Wiener et Fonctions Aléatoires Gaussiennes. X, 383 pages. 1974. DM 32,–

Vol. 380: M. Petrich, Rings and Semigroups. VIII, 182 pages. 1974. DM 18,–

Vol. 381: Séminaire de Probabilités VIII. Edité par P. A. Meyer. IX, 354 pages. 1974. DM 32,–

Vol. 382: J. H. van Lint, Combinatorial Theory Seminar Eindhoven University of Technology. VI, 131 pages. 1974. DM 18,–

Vol. 383: Séminaire Bourbaki – vol 1972/73 Exposés 418-435 IV, 334 pages. 1974. DM 30,–

Vol. 384: Functional Analysis and Applications, Proceedings 1972. Edited by L. Nachbin. V, 270 pages. 1974. DM 22,–

Vol. 385: J. Douglas Jr. and T. Dupont, Collocation Methods for Parabolic Equations in a Single Space Variable (Based on C^1-Piecewise-Polynomial Spaces). V, 147 pages. 1974. DM 16,–

Vol. 386: J. Tits, Buildings of Spherical Type and Finite BN-Pairs. IX, 299 pages. 1974. DM 24,–

Vol. 387: C. P. Bruter, Eléments de la Théorie des Matroïdes. V, 138 pages. 1974 DM 18,–

Vol. 388: R. L. Lipsman, Group Representations. X, 166 pages. 1974. DM 20,–

Vol. 389: M.-A. Knus et M. Ojanguren, Théorie de la Descente et Algèbres d' Azumaya. IV, 163 pages. 1974. DM 20,–

Vol. 390: P. A. Meyer, P. Priouret et F. Spitzer, Ecole d'Eté de Probabilités de Saint-Flour III – 1973. Edité par A. Badrikian et P.-L. Hennequin. VIII, 189 pages. 1974. DM 20,–

Vol. 391: J. Gray, Formal Category Theory: Adjointness for 2-Categories. XII, 282 pages. 1974. DM 24,–

Vol. 392: Géométrie Différentielle, Colloque, Santiago de Compostela, Espagne 1972. Edité par E. Vidal. VI, 225 pages. 1974. DM 20,–

Vol. 393: G. Wassermann, Stability of Unfoldings. IX, 164 pages. 1974. DM 20,–

Vol. 394: W. M. Patterson 3rd, Iterative Methods for the Solution of a Linear Operator Equation in Hilbert Space – A Survey. III, 183 pages. 1974. DM 20,–

Vol. 395: Numerische Behandlung nichtlinearer Integrodifferential- und Differentialgleichungen. Tagung 1973. Herausgegeben von R. Ansorge und W. Törnig. VII, 313 Seiten. 1974. DM 28,–

Vol. 396: K. H. Hofmann, M. Mislove and A. Stralka, The Pontryagin Duality of Compact O-Dimensional Semilattices and its Applications. XVI, 122 pages. 1974. DM 18,–

Vol. 397: T. Yamada, The Schur Subgroup of the Brauer Group V, 159 pages. 1974. DM 18,–

Vol. 398: Théories de l'Information, Actes des Rencontres de Marseille-Luminy, 1973. Edité par J. Kampé de Fériet et C. Picard XII, 201 pages. 1974. DM 23,–

Vol. 399: Functional Analysis and its Applications, Proceedings 1973. Edited by H. G. Garnir, K. R. Unni and J. H. Williamson. XVII, 569 pages. 1974. DM 44,–

Vol. 400: A Crash Course on Kleinian Groups – San Francisco 1974. Edited by L. Bers and I. Kra. VII, 130 pages. 1974. DM 18,–

Vol. 401: F. Atiyah, Elliptic Operators and Compact Groups. V, 93 pages 1974. DM 18,–

Vol. 402: M. Waldschmidt, Nombres Transcendants. VIII, 277 pages. 1974. DM 25,–

Vol. 403: Combinatorial Mathematics – Proceedings 1972. Edited by D. A. Holton VIII, 148 pages. 1974. DM 18,–

Vol. 404: Théorie du Potentiel et Analyse Harmonique. Edité par J. Faraut. V, 245 pages. 1974. DM 25,–

Vol. 405: K. Devlin and H. Johnsbråten, The Souslin Problem. VIII, 132 pages. 1974. DM 18,–

Vol 406: Graphs and Combinatorics – Proceedings 1973. Edited by R. A. Bari and F. Harary. VIII, 355 pages. 1974. DM 30,–

Vol. 407: P. Berthelot, Cohomologie Cristalline des Schémas de Caracteristique p > o. VIII, 598 pages. 1974. DM 44,–

Vol. 408: J Wermer, Potential Theory. VIII, 146 pages. 1974. DM 18,–

Vol. 409: Fonctions de Plusieurs Variables Complexes, Séminaire François Norguet 1970–1973. XIII, 612 pages. 1974. DM 47,–

Vol. 410: Séminaire Pierre Lelong (Analyse) Année 1972–1973. VI, 181 pages. 1974. DM 18,–

Vol 411: Hypergraph Seminar. Ohio State University, 1972. Edited by C Berge and D. Ray-Chaudhuri. IX, 287 pages. 1974. DM 28,–

Vol. 412: Classification of Algebraic Varieties and Compact Complex Manifolds. Proceedings 1974. Edited by H. Popp. V, 333 pages. 1974 DM 30,–

Vol. 413: M Bruneau, Variation Totale d'une Fonction. XIV, 332 pages. 1974. DM 30,–

Vol. 414: T Kambayashi, M. Miyanishi and M Takeuchi, Unipotent Algebraic Groups VI, 165 pages 1974 DM 20,–

Vol. 415: Ordinary and Partial Differential Equations, Proceedings of the Conference held at Dundee, 1974. XVII, 447 pages. 1974. DM 37,–

Vol. 416: M. E. Taylor, Pseudo Differential Operators. IV, 155 pages. 1974. DM 18,–

Vol. 417: H. H. Keller, Differential Calculus in Locally Convex Spaces. XVI, 131 pages. 1974. DM 18,–

Vol 418: Localization in Group Theory and Homotopy Theory and Related Topics Battelle Seattle 1974 Seminar. Edited by P J. Hilton. VI, 171 pages. 1974. DM 20,–

Vol. 419: Topics in Analysis – Proceedings 1970. Edited by O E. Lehto, I. S. Louhivaara, and R. H. Nevanlinna. XIII, 391 pages. 1974. DM 35,–

Vol. 420: Category Seminar Proceedings, Sydney Category Theory Seminar 1972/73. Edited by G. M Kelly VI, 375 pages. 1974 DM 32,–

Vol 421: V Poénaru, Groupes Discrets. VI, 216 pages. 1974. DM 23,–

Vol 422: J.-M. Lemaire, Algèbres Connexes et Homologie des Espaces de Lacets. XIV, 133 pages. 1974. DM 23,–

Vol. 423: S. S Abhyankar and A. M. Sathaye, Geometric Theory of Algebraic Space Curves. XIV, 302 pages. 1974. DM 28,–

Vol 424: L. Weiss and J Wolfowitz, Maximum Probability Estimators and Related Topics. V, 106 pages. 1974. DM 18,–

Vol. 425: P. R. Chernoff and J. E. Marsden, Properties of Infinite Dimensional Hamiltonian Systems. IV, 160 pages. 1974. DM 20,–

Vol. 426: M. L. Silverstein, Symmetric Markov Processes. IX, 287 pages. 1974. DM 28,–

Vol 427: H. Omori, Infinite Dimensional Lie Transformation Groups XII, 149 pages 1974 DM 18,–

Vol. 428: Algebraic and Geometrical Methods in Topology, Proceedings 1973. Edited by L. F. McAuley. XI, 280 pages. 1974. DM 28,–